AUSTRALIAN DEFENCE AND DETERRENCE

A 2024 Update

ROBBIN LAIRD

This book is dedicated to the memory of Air Marshal Vale Errol (Mac) McCormack

This is what was published by the Williams Foundation after his death shortly before our 11 April 2024 seminar:

"Errol had a distinguished career in the RAAF, commanding at squadron, wing and Air Force levels and retired from the RAAF as Chief of Air Force in May 2001. After retiring he was a Board Member and consulted for many local and foreign companies dealing with Defence.

"AIRMSHL McCormack was the inaugural Chair of the Sir Richard Williams Foundation and was instrumental in founding the organisation."

Library of Congress Control Number: 2024910906

The cover photo is credited to Dreamtime. Illustration 209804403 © Tal Hayoun

Upside down world map silhouette in black. A revered upside down, south up map of the world - a 180 degree rotation of the map which is much better way to understand Australia and why a maritime strategy is so important to the country.

Contents

Introduction

I have been coming to Australia since 2014 to participate in and write the seminar reports for the Sir Richard Williams Foundation.

What started out as a Foundation focusing primarily on airpower has transitioned over that period to focusing on the Australian Defence Force (ADF) as an integrated force, and one increasingly focused on the Indo-Pacific region.

The latest Sir Richard Williams Foundation.report focused on the core conference theme which was entitled: *The Multi-Domain Requirements of an Australian Defence Strategy.* I have included this report in the current book.

But unique in my work has been providing a detailed look at the ADF transition since 2014, which I have provided in a series of books which I have published over the past few years.

I am the only American who is a fellow with the Foundation, and find the recent surge of American expertise on Australia associated with AUKUS to be of some concern.

AUKUS is a Rorschach test. You can see in it what you want. It is loaded with ambiguity, the kind that can lead to serious conflict in the United States, the UK and in Australia about who is supposed to do what.

Our alliance is too important to be reduced to chanting AUKUS, rather than frankly discussing our common interests and our differences. And the domestic divergencies in our societies is very significant and has clear implications for the future. An AUKUS totem doesn't solve this.

The book starts with a seminal essay by my friends and colleagues John Blackburn and Anne Borzycki. Their essay addresses the challenges which the political system in Australia poses to having an effective defence policy. Australia clearly needs a national defence and security strategy and not just the AUKUS totem.

This essay was originally written for Murielle Delaporte's publication *Operationnels* and a lead article in her latest journal publication prepared for the Eurosatory 2024 Conference, held in Paris in June 2024. John is a member of the editorial board of the journal and wrote this essay along with his colleague Anne Borzycki for that publication. We thank the editor and the authors for permission to include an edited version in this book.[1]

I next have included my checklist of key challenges which need to be addressed in building such a strategy. This checklist was published last year in my book entitled *Australia and Indo-Pacific Defence: Anchoring a Way Ahead.*

I then turn to the 11 April 2024 seminar report, which was a real challenge to write for the government's future force ambitions are being paid for by cutting current ADF capabilities. Not only does this put Australia into a difficult situation, but I am very skeptical based on many years of experience of how good government's are at projecting future force requirements and their ability to know the future in any case.

How good did we do in forecasting 2020 in 2019?

I have included next the interviews which I conducted in April 2024 before and after the seminar which provide a rich tapestry of assessments of the challenges for Australian defence by Australians who are well positioned to comment on these challenges.

Each interview was published on either *Second Line of Defense* or *Defense Information* and most can be found on The Richard Williams

Foundation website as well. The date indicated after the title of the interview was the date it was published on the website.

This is a journey and only future history will tell us definitively where we were in that journey in 2024.

Dr. Robbin F. Laird

Research Fellow

Sir Richard Williams Foundation

ONE

The Impact of the Australian Political System on National Security

By John Blackburn and Anne Borzycki

Australia has a lengthy history of seeking protection from great and powerful western saviours due to a long-held fear of invasion from "the north."

Stunned by the withdrawal of British Forces following the surrender of Singapore in WWII, it turned its attention to another large and powerful ally, the United States.

This dependence grew stronger following the UK's withdrawal from 'East of Suez,' which was precipitated by the Suez Crisis in the late 1950s. The domino theory of the 1960s, which predicted the spread of communism in South-east Asia, fuelled Australian public concern and was used for political gain.

Australia relied on the United States as its security guarantor, initially through the 1951 ANZUS treaty. A treaty that, with the rupture of relations between the United States and New Zealand, no longer exists in full force but is still trumpeted by Australian politicians whenever it is judged beneficial.

The bilateral U.S.-Australia alliance is the most strategically significant relationship, despite the announcement of the AUKUS partnership in September 2023 which implied some degree of relevance for the UK in the region.

National security is based on capabilities, competence, and partnerships; complacency is a major hindrance to that security.

Is Australia, in fact, secure in 2024, or is Australia a complacent nation adrift in the South Pacific?

Politics, Geopolitics and the Defence of Australia - The 21st Century Context

On 14 September 2001, at a press conference at the Parliament House of Australia, Prime Minister John Howard invoked the Australia, New Zealand, and United States Security Treaty (ANZUS) for the first and only time, following the terrorist attacks on 11 September 2001 in the United States.[1]

John Howard had always been a conventional, if somewhat conservative, supporter of the Australian-U.S. alliance. Neither a champion nor a critic. That would all change after 9/11.

In 2006 Professor Emeritus Robert Manne wrote that "from this moment Howard worked conscientiously to create a new vision of the future: of an Australia deeply integrated – strategically, economically, socially and culturally – into the most formidable empire the world has ever seen … the U.S.."[2]

Possibly a slightly hyperbolic assessment in 2006, but one that ultimately became Australia's reality.

Prime Minister Kevin Rudd's *Plan for Defence*, released during the election campaign of 2007, noted that the Labor approach to Defence "… is built around three fundamental pillars – [first] Australia's alliance with the United States … [which] is fundamental to our national security and our long-term strategic interests."[3]

At the launch of the Defence White Paper (DWP 2009) on 2 May 2009, Kevin Rudd stressed that Australia's alliance with the United States would remain the bedrock on which the country's national security is built: "The document judges that the United States will remain the most powerful and influential military actor out to 2030."[4]

The DWP 2009 announced the need to upgrade and enhance maritime capabilities as a consequence of an assessment of the changing strategic environment.

Specifically, "by the mid-2030s, we will have a heavier and more potent maritime force. In the case of the submarine force, the Government takes the view that our future strategic circumstances necessitate a substantially expanded submarine fleet of 12 boats in order to sustain a force at sea large enough in a crisis or conflict to be able to defend our approaches (including at considerable distance from Australia, if necessary), protect and support other ADF assets, and undertake certain strategic missions where the stealth and other operating characteristics of highly-capable advanced submarines would be crucial."[5]

The phrase "*to sustain a force at sea large enough in a crisis or conflict to be able to defend our approaches (including at great distance from Australia, if necessary*" outlines a picture for the future submarine operating far from home, protecting our sea lanes and contributing to Asian regional security, and an unidentified threat, In 2009 Australia's political leadership was hesitant and discreet in overtly naming 'China' as the target of defence planning and capability expansions.

Over the period in which the Australian Labor Party (ALP) was in government (2007 – 2013), despite the internal fractures of two leadership coups, the commitment of Australian Defence Force (ADF) personnel to both the Middle East and Afghanistan continued almost unquestioned. The relationship with the U.S. remained foundational, and indeed, deepened.

In 2011, then Prime Minister Julia Gillard and President Barack Obama, announced two new force posture initiatives that would significantly enhance defence cooperation between Australia and the United States.

The media release said that "Starting next year, Australia will welcome the deployment of U.S. Marines to Darwin and Northern Australia, for around six months at a time, where they will conduct exercises and training on a rotational basis with the Australian Defence Force … The intent in the coming years is to establish a rotational presence of up to a 2,500-person Marine Air Ground Task Force."[6]

The leaders also agreed to closer cooperation between the Royal Australian Air Force and the U.S. Air Force that would result in increased rotations of U.S. aircraft through northern Australia.

The announcement prompted rapid, and at times, strident, criticism from the Chinese government. *The New York Times* noted that the deployment / basing of U.S. Marines in Australia "prompted a sharp response from Beijing, which accused Mr. Obama of escalating military tensions in the region …"[7]

In May 2013, just months before losing the federal election in September, the ALP released the Defence White Paper 2013.

This White Paper saw a significant shift in the language used in reference to China. From being regarded as a growing economic power in the region with whom meaningful dialogue and engagement was important, in conjunction with allies and friends, we now had a comforting paragraph reassuring Australians that there was no need to choose between the U.S. and China.

Australia's national security and economic reality appeared to be well balanced between the strategic rivals with whom we were connected.[8]

When Labor lost power to the Liberal/National Party coalition in September 2013, Tony Abbott, the new Prime Minister, took over the submarine program with a few proposals of his own. Prime Minister Abbott became infamous for his 'captain's pick' moments in which he made decisions without consultation with colleagues, advisers, experts or indeed, anyone.

Abbott became convinced that the Japanese Soryu-class submarine was precisely what Australia needed. Media reporting during 2014 widely reflected this position.[9] Local shipbuilding businesses were angered at the probable loss of jobs and commercial opportunities.

In early 2014, the newly elected Abbot Liberal/National Party government actively sought to deepen economic ties with China through a free-trade agreement, at the same time as balancing strong criticism of Chinese actions in the East China Sea.[10]

Prime Minister Tony Abbott famously described the relationship with China at that time as one based on 'fear and greed'.[11] Fear of China's growth, wealth and ambitions, but also greedy for the trade this growth enabled and the economic benefits that flowed to Australia. A variation of the 'we don't need to choose sides' position of the previous ALP government.

By February 2015, following a series of internal party tensions and leadership threats and challenges, Prime Minister Abbott announced that a Competitive Evaluation Process would be undertaken as part of the Future Submarine acquisition strategy. This evaluation process involved France (DCNS), Germany (ThyssenKrupp Marine Systems—TKMS) and the Government of Japan as potential international design partners.[12]

In September 2015, Prime Minister Turnbull overthrew Tony Abbott, in yet another bitter internal party coup. The free-trade agreement, sought by Abbott, was finally signed in December 2015, and the praise was heaped upon President Xi by Australian politicians at a formal dinner to mark the occasion in the Great Hall at Parliament House Canberra.[13]

Not long after the signing of the Free Trade Agreement, the Government released Defence White Paper 2016 (DWP 2016) which acknowledged "The relationship between the United States and China is likely to be characterised by a mixture of cooperation and competition depending on where and how their interests intersect."[14]

The DWP 2016 verified that the Future Submarine Program would be a rolling acquisition program.

On 26 April 2016 Prime Minister Malcolm Turnbull declared DCNS (now Naval Group) of France as the preferred international design partner for Australia's future submarine. The successful design was the Shortfin Barracuda Block 1A conventional submarine, which was based on the French Barracuda nuclear powered submarine.

Australia had evidently not learned the lessons from prior defence acquisitions which involved taking a proven design by a foreign manufacturer, and then extensively changing it to meet perceived special Australian requirements.

On 30 September 2016 a contract worth between $450 and $500 million was signed between the Australian Government and Naval Group for the 'design and mobilisation' of Australia's 12 Future Submarines. This included the development of Adelaide's Osborne North submarine facility for this purpose.[15]

Regionally, the honeymoon period that followed the free trade agreement with China did not last long. Prime Minister Turnbull ignited the ire of the Chinese government in December 2017 when the federal government passed the Foreign Influence Transparency Scheme against the backdrop of a series of high-profile scandals involving CCP influence in Australian politics.

Though the law itself was not targeted at any one country in particular, Prime Minister Turnbull's use of the politically-charged phrase "the Australian people stand up," and his open criticism of CCP influence in Australian politics generated diplomatic blowback in Beijing.[16]

Then, in August 2018, Australia prohibited Chinese giants Huawei and ZTE from participating in building Australia's 5G networks. A tit-for-tat escalation of trade actions/sanctions followed which only worsened after April 2020 when the new Prime Minister, Scott Morrison (who overthrew Malcolm Turnbull in yet another bitter internal party coup), endorsed an independent investigation into the origins of the COVID-19 pandemic, infuriating the Chinese officials.

The 2020 Defence Strategic Update (DSU) signed by both the Prime Minister and Minister for Defence, gave no hint of what was to come for the submarine project, just one year later.

The problems with the submarine project were a poorly kept secret around Canberra – within Defence and amongst successive Prime Ministers, Ministers, and industry. Indeed, there was minimal reference to the project in the DSU; merely a brief mention that the "Government will continue to deliver this nationally significant program of investment in ships and submarines…"[17]

And what was said about China?

Yet another step-change in the language used, for example: "While still unlikely, the prospect of high-intensity military conflict in the Indo-Pacific is less remote than at the time of the 2016 Defence White Paper, including high-intensity military conflict between the United States and China ..."[18]

Then Prime Minister Morrison further worsened international relationships when, in September 2021, he announced the cancellation of the French submarine project after entering into a new, secretly negotiated, agreement with the United Kingdom and the United States.

With a single stroke Morrison aggrieved both China and France.

The French Foreign Minister Jean-Yves Le Drian called the trilateral submarine deal evidence of 'duplicity', 'treachery' and a 'stab in the back.' He also criticised Morrison's lack of candour. France recalled its ambassadors to the United States and Australia.[19] French President Emmanuel Macron said Prime Minister Morrison lied to him about his intentions with the deal, and that his trust in Australia had been deeply damaged.

This came a day after U.S. President Joe Biden expressed deep concerns about the handling of the secret plan to dump Naval Group from the future submarine project, labelling it as "clumsy" and "not done with a lot of grace".[20]

Many Australians would have a similar opinion of Morrison's honesty and clumsiness, especially after many of his activities as Prime Minister were made public following his defeat in the 2022 federal election. Of course, many Australians would also have vivid memories of France's less-than-trustworthy actions in the Pacific region over previous decades.[21]

The Emergence of AUKUS

AUKUS, the trilateral security partnership for the Indo-Pacific region between Australia, the United Kingdom (UK), and the United States (U.S.) was announced on 15 September 2021.

The decision by then Prime Minister Morrison, was a complete surprise to the Australian people. In the March 2023 *Australia in the*

World podcast, the late Alan Gyngell, former Director-General of the Office of National Assessments, a leading Australian foreign policy adviser and academic, was quoted as saying that "the most surprising thing about this announcement of the largest project ever undertaken by the Commonwealth of Australia remains the fact that there has been no formal articulation of the reasons for the decision. No report, no speech to parliament, no speech at all, other than the sales patter from successive governments: 'China is more assertive, the rules-¬based order is under threat, nuclear submarines are just what Australia needs."[22]

It's as if AUKUS, once proclaimed by the Liberal Prime Minister, became sacrosanct, with the then-opposition leader, now Labor Prime Minister, Albanese agreeing to the collaboration in just hours. There has been no political debate on this critical matter for Australia's security.

Now, it appears to be nearly treasonous to question the decision's merits; citizens debating the subject were effectively mocked by the Chief of Navy, Vice Admiral Hammond, who urged Australians to ignore 'hand-wringing' doubters of the AUKUS pact.[23]

Have political talking points spread to the highest levels of the Australian Defence Force (ADF)?

Whilst the 'U.S.' part of AUKUS could be seen as a logical fit, the value of the 'UK' part was more difficult to justify. Assuming that AUKUS planning was already underway, and was a little over a year from being announced, what did the 2020 Strategic Update say about Australia's relationship with the United Kingdom?

In the 68-page document there was not one mention of the United Kingdom. Yet by September 2021, Prime Minister Morrison was quoted as describing AUKUS as a "forever partnership for a new time between the oldest and most trusted of friends."[24]

Recently, the Australian Defence Minister that the "UK has a much greater presence in the Indo-Pacific than we have seen in a very long time".[25]

But apparently this is news in Britain. The UK Parliament's *UK Defence and the Indo-Pacific – Report Summary*, dated 24 October 2023, stated: "Although we welcome the progress made in the region, we reject the notion that the 'tilt' has been 'achieved' from a Defence perspective.

"With only a modest presence compared to allies, little to no fighting force in the region, and little by way of regular activity, UK Defence's tilt to the Indo-Pacific is far from being achieved. If we aspire to play any significant role in the Indo-Pacific this would need a major commitment of cash, equipment and personnel, or potentially rebalancing existing resources. The UK Government's future strategy for the Indo-Pacific is still unclear."[26]

So much for a "forever partnership" with the UK in the region.

Political commentators, national security experts, academics and the media in general were still trying to come to terms with what AUKUS practically meant for Australia when, in April 2022, Prime Minister Morrison announced that a federal election would take place in May 2022.

The Liberal/National Party coalition had styled themselves as the bastions of security and the 'liberal' ALP were frequently derided as 'soft.' In the period leading up to the election, the ALP had little political wiggle-room to change stance on AUKUS or risk the usual 'soft on security' barbs being levelled at them.

At the end of the day, after nearly a decade of 'conservative' government, the Liberal/National Party was defeated, and the ALP was returned to government in May 2022 – with AUKUS steadfastly a part of their policy platform as well.

To emphasise this reluctance to 'rock-the-boat' so to say, the Australian public policy journal, *Pearls and Irritations*, observed: "It is understandable that the leadership team heading towards an election was determined not to be wedged on defence policy ... [but] why would an intelligent forward-looking government fail to fully assess the implications of secret international negotiations made by an unreliable former prime minister in the dying days of a discredited government?

"Why is our apparently very capable leadership team so timid about questioning the terms and range of preconditions for the AUKUS agreement?"[27]

AUKUS firmly intact, the new Labor government, under the leadership of Prime Minister Albanese, made immediate attempts to repair / renew the relationship with China – primarily as it related to trade.

There was an urgent political and economic need to get the export market back on track. And there were early breakthroughs in late 2022 as both the Foreign and Trade Ministers were able to meet with Chinese government officials to discuss avenues for lifting sanctions / tariffs and reopening channels of communication.[28]

Just one month after the election and during his visit to India, the Defence Minister/Deputy Prime Minister said in an interview with Australia's ABC: "… for India and Australia, China is our largest trading partner. For India and Australia, China is our biggest security anxiety. We're both trying to reconcile those things, which is not an easy problem to solve."[29]

Australia's perennial paradox, and a juggling act that is only getting more challenging. Australia's defence policy at a macro level rarely moves too dramatically from one government to the next, nonetheless, within the first 100 days of office, the Albanese government commissioned the independent Defence Strategic Review (DSR) 2023.

Within the body of the DSR, the language regarding China assumes a more ardent tone: "Intense China-United States competition is the defining feature of our region and our time. Major power competition in our region has the potential to threaten our interests, including the potential for conflict ... China's military build-up is now the largest and most ambitious of any country since the end of the Second World War …

"This build-up is occurring without transparency or reassurance to the Indo-Pacific region of China's strategic intent …China's assertion of sovereignty over the South China Sea threatens the global rules-based order in the Indo-Pacific in a way that adversely impacts Australia's national interests. China is also engaged in strategic competition in Australia's near neighbourhood."[30]

The relationship with China is a genuinely baffling example of cognitive dissonance. According to the Department of Foreign

Affairs and Trade, China is Australia's largest two-way trading partner, accounting for 26 per cent of our goods and services trade with the world in Financial Year 2022-23.[31] Yet China remains the significant factor in Australia's defence planning, and posturing.

In the years since John Howard committed Australia to the United States, the relationship has been turbocharged to the point where the 'dependency' (rather than interdependence) could impact Australia's independence and sovereignty.

The DSR 2023 reflects this turbocharging: "Our Alliance with the United States will remain central to Australia's security and strategy. The United States will become even more important in the coming decades. Defence should pursue greater advanced scientific, technological, and industrial cooperation in the Alliance, as well as increased United States rotational force posture in Australia, including with submarines. Contrary to some public analysis, our Alliance with the United States is becoming even more important to Australia."[32]

There is an expression that was coined during the Cold War, by then-Prime Minister Menzies, to define what was virtually a canonical faith in Australia's international outlook – that Britain and the U.S. formed a protective umbrella for the country in a threatening Asia.[33]

The 'great and powerful friends' would be the protectors. From Federation until World War 2 it was the UK, and since then, the U.S.

- And while solid, reliable alliances are fundamental to global and regional stability, at what point does an 'alliance' become an 'addiction', an unhealthy dependency?
- Has this historic reliance weakened Australia?
- Has it given rise to complacency?
- Has it stifled creativity and innovation?
- Has it denied the nation a shared narrative?
- Has it created a gilded cage from which there is no escape?

With AUKUS, Australia has become a loyal partner to two empires – one in an apparent political civil war and the other of little relevance to the Asia Pacific region, according to its own Parliament, as discussed previously.

Indeed, the partnership may, in the end, constrain sovereign thinking, capability and sovereignty given the degree of compliance required with U.S. International Traffic of Arms Regulations and the Australian Governments inventive definition of 'sovereign' in the Defence Industry Policy.

The Policy defines foreign primes that have operations in Australia as 'Australian' and as 'sovereign' as actual Australian companies, which have majority Australian ownership and controls and headquarters in Australia.[34]

The Current Approach

On 24 April 2023 the Australian Labor Government released the Defence Strategic Review (DSR), which examined whether Australia possessed the required defence capabilities, posture, and preparation to best defend itself and its interests in the current strategic environment.

The DSR concluded that the ADF was "not fully fit for purpose" in the face of escalating security challenges. This occurred despite an investment of around A$695 billion in defence from 2007 to 2022 by both the Labor and Liberal Coalition governments.

How could this have occurred? What does the present Labor government plan to do about it?

The DSR stated that "Australia's strategic circumstances and the risks we face are now radically different - the U.S. is no longer the unipolar leader of the Indo-Pacific. Major power competition in the region has the potential to threaten our interests, including the potential for conflict. China's military build-up is now the largest and most ambitious of any country since the end of WW2 … This necessitates a managed, but nevertheless focused, sense of urgency. It is clear that a business-as-usual approach is not appropriate."[35]

However, Australia's strategic situation is more complex than depicted in the DSR. Australians are currently dealing with concurrent, and in some cases existential, concerns.

These include climate change and the urgent need to reduce emissions, rising global and regional security risks, a global pandemic with long-term societal and economic consequences, a global energy transformation in which we lag behind the developed world, and a global market model that has resulted in reduced resilience due to a lowest price, just-in-time philosophy.

Increased levels of internal instability, societal unrest, and a lack of trust in government and institutions among our AUKUS partners raise concerns about the possibility of similar tendencies occurring in Australia.

In 2021, our Institute of Integrated Economic Research – Australia produced a report on Australia's national resilience titled *Australia – A Complacent Nation … Our reactions are too little, too late, and too short-sighted.*[36]

It highlighted Australia's lack of resilience in the face of current emerging threats and risks. Little progress has been made in the intervening three years.

Whilst the DSR did identify that a central component of deterrence for national defence is resilience, listing 12 examples of critical resilience requirements, there is no indication of how Government intends to address them; listing them is not acting.

In each case there needs to be a risk assessment before a strategy and plan can be developed. For example, given national concerns regarding the state of our energy systems and massive import dependencies, it is sobering to realise that the last time Australia had a National Energy Security Assessment, was in 2011. There appears to be no coherent plan.

The DSR was a largely bottom-up, platform-focussed review and plan, delivered incrementally over a period of 10 months, following its initial announcement, and devoid of any National Security Strategy or Defence Strategy to provide context and guidance. A Defence Strategy was delivered in April 2024, but there is still no elusive National Security Strategy.

It is also difficult to discern how the DSR has taken account of the numerous failures to implement previous reviews, as well as what lessons were learned from the actions that were only partially implemented.

The Government has stated that "Realising the ambition of the Review will require a whole-of-government effort, coupled with significant financial commitment and major reform."[37]

Despite the aspirational phrasing and free use of the terms "soon" and "sooner" in numerous political statements, there is no significant new financing for DSR implementation this decade.

The Australian Defence Department has often struggled to manage complicated change programmes, but now they are expected to succeed with this large and under-resourced reform programme. Reorganisations, reallocation of resources and cancellation / changes in projects are assumed to be all that is necessary in the near term.

The failure to resource the DSR adequately could mean that our deployable military operational capability will be less at the end of this decade than it is today and this right when the threat envisaged by the review could appear.

Since 2000 there have been six Chiefs of Defence who have been directed by 13 Defence Ministers (two of whom only served 271 and 272 days respectively), 13 Assistant Ministers, and 12 Ministers for Defence Industry, Defence Materiel, and Defence Personnel. A parade of 38 Defence related Ministers over 23-years, each with a party-political agenda, and often with little real understanding of National Security or Defence.

This august group tasked a total nine Defence related White Papers, and nine Defence reform, restructure, review, and transformation programs. Were any lessons learned during this 24-year journey? Apparently very little, given the findings of the 2023 DSR.

The Submarine Saga

In 2009 the Australian Government announced that by the mid-2030s, we would have a heavier and more potent maritime force. Our future strategic circumstances necessitated a substantially expanded submarine fleet of 12 boats in order to sustain a force at sea large enough in a crisis or conflict.

Over the intervening 15 years, successive Governments have identified Japanese, then French and finally a mixed fleet of used U.S. submarines, and a yet to be designed UK submarine type with delivery of the latter to occur in the 2040s.

This, a decade after they are needed according to DWP 2009, and a decade after the heightened threat of conflict in the region could materialise, according to the DSR 2023.

Of particular interest, given growing public scepticism regarding the actual feasibility of the AUKUS submarine plans, is the view of the current opposition leader, Peter Dutton, who was the Defence Minister at the time of the AUKUS announcement in 2021.

In March 2023, Dutton warned against acquiring a future submarine fleet from the UK, stating that advice he received as defence minister before the last election_was that the British companies involved in the UK's submarine production had no extra capacity to support an Australian program and insisting that American Virginia-class submarines remained the best and most practical option for Australia.[38]

A pity it took him 18 months to speak up but that's politics.

This incredible US/UK hybrid submarine fleet package is thoroughly examined in Professor Hugh White's superb essay "Dead in the Water," which was published in the *Australian Foreign Affairs Journal* in February 2024.[39]

He explains the decision's assumptions, the significant likelihood that the plan will be disrupted, and how deploying the proposed nuclear submarines in the future will have little or no impact the military balance in Asia or on Australia's security.

He discussed the capacity differences between conventional and nuclear-powered submarines, stating that while a nuclear-powered submarine may be equivalent to two conventional boats, it costs five times as much.

Professor White summed up the AUKUS deal as follows: "AUKUS also gave Morrison a high-profile defence initiative which provided a national security focus for the 2022 election campaign, and presented a chance to wedge Labor if they opposed it. Labor dodged that bullet by fully supporting AUKUS, partly from a pathological fear of a political brawl with the Coalition on any national security issue, but also because many of the party's leaders believe as fervently in supporting America against China as the Coalition does.

"It is a bipartisanship built upon political opportunism and strategic complacency on both sides ... the sheer scale of AUKUS puts it in a class of its own as an exemplar of bureaucratic incompetence."[40]

We will leave the technical and capability arguments to Professor White and others. But one point which we will highlight, based on our military experience over many decades, is that the transition complexity for completing a life-of-type-extension on the existing Collins Class boats whilst managing the introduction of around three used U.S. Virginia Class nuclear boats (with their reported maintenance / availability issues) whilst continuing to manage a problematical surface fleet ship construction program and concurrently introducing a yet to be designed UK Nuclear Boat by a Navy that is already 6% below authorised staff levels and is experiencing recruiting difficulties is a recipe for disaster.[41]

Given Australia's Defence procurement history, the possibility of this occurring on schedule and on budget is close to zero. However, this will surely provide excellent lessons to be ignored in a future Defence review!

The Impact of The Australian Political System on National Security

Given our previous assessment of the Australian political system's impact on defence, we will now look at the impact on national security, a much bigger canvas for politicians to muddy.

Our first concern is that the analysis of our national security continues to be conducted through a somewhat narrow military platform lens.

For example, the commitment to what appears to a 25-year plus program to acquire two different types of nuclear submarines from two separate nations, for a cost in the order of $350B is being justified with little opportunity for the Australian people to understand why the decision was taken.

This is particularly problematic considering that Australia lacks a national risk assessment, a national security strategy, and a national defence strategy. The latter, however, was delivered in early 2024, three years after the AUKUS decision was announced.

All of this has transpired in a highly complicated security context, with uncertainty in global economic systems already generating palpitations in many countries. This is the result of a broken political system over the preceding decades, which repeatedly fails to learn from previous mistakes and failures.

This situation has been described by former Liberal Party opposition leader John Hewson as follows: "Voters are also increasingly tired of the adversarial nature of the contest between Liberal and Labor – pointless short-term point scoring unrelated to either the national interest or to addressing important issues. The main parties are too often focused on a power game played among themselves, saying, or doing whatever they believe is necessary to win, where the endgame has little to do with listening to voter concerns or providing better government."[42]

Our second concern is the oft venerated ANZUS Treaty; the negotiation of ANZUS in 1951 was the first instance of Australia forming a political alliance without the involvement of Britain, causing some tension with 'the motherland'.

Australia's commitment to U.S.-led causes such as the Vietnam War, while not formalised under ANZUS, have been linked to the treaty. Many political commentators suggested that Australia's strong involvement in Vietnam was a means of proving our usefulness to the alliance, demonstrating Australian loyalty should we need U.S. support in the future.

Despite being in operation for more than 60 years, the ANZUS treaty has only been formally invoked once in 2001 as a response to the September 11 terrorist attacks in New York and Washington, leading to Australia's long-term involvement in the United States led 'War on Terror'.

The ANZUS treaty has occasionally experienced difficulties. Australia expressed disappointment at the minimal support received from the United States during the Konfrontasi conflict in Indonesia and Malaysia in the early 1960s.

There has also been ongoing discussion about the contemporary relevance of an alliance that was created in a 1950s Cold War environment. Since the early 1980s the United States has effectively absolved itself of any obligation to New Zealand when the latter became a nuclear-free country and refused entry to its ports by potentially nuclear-capable US warships.

Today, Australia maintains bilateral security relationships with both New Zealand and the United States under the treaty's banner. However, while the treaty has not been formally revoked after the breakdown of relations between the United States and New Zealand, the treaty between the three nations no longer fully exists in practice.[43]

Our third concern is the cost, past, present, and ongoing, of our political decisions. The dismal and costly results of Australia's involvement in the Vietnam, Iraq and Afghanistan wars is the result of repeated failures of political leaders of many nations and the inability to conceive of any alternative to address persistent security concerns.

The authors have worked closely with the U.S. armed forces and have respect and admiration for the men and women of the U.S. Services, but less for the parade of politicians from both the U.S. and Australia.

So, what was the cost of Australia's commitment to the U.S. Alliance?

- Vietnam - Australia suffered 521 deaths and a little over 3,000 wounded during the Vietnam War. 60,000 Australian soldiers were sent to the front, with a third of them being conscripts. The costs associated with participation were substantial, in the order of $220M, including both the direct costs of deployment and support for South Vietnam.[44]

- Iraq – Around 2000 ADF personnel served in this conflict. No Australian military personnel were killed in direct combat action during Operation Falconer or Operation Catalyst; however, one Australian soldier died in 2015 as a direct result of injuries sustained in an IED blast in 2004. The cost of the Iraq war to Australian taxpayers is estimated to have exceeded A$5 billion.[45]
- Afghanistan - the Afghanistan conflict resulted in 41 Australian military deaths, significant injuries, and a financial cost exceeding $10 billion; around 30,000 ADF personnel served over two decades in Afghanistan.[46]

What is harder to estimate is the ongoing physical and mental health impacts on the men and women who served in these wars. When compared to the wider Australian population, the age-adjusted rate of suicide was 26% higher for ex-serving males, and 107% higher for ex-serving females. The rate for ex-serving males who separate involuntarily for medical reasons is over three times higher than those who separated voluntarily.[47]

Even to this day, the Government, Defence, and the Department of Veterans' Affairs have been unable to address the breadth and depth of concerns in the veteran community.

This list also does not account for the millions of civilians who died as a result of these conflicts.

The loss of the Vietnam war, the eventual return of the Taliban to power in Afghanistan after a chaotic withdrawal of U.S.-led coalition forces, and the confused security situation following the Iraq wars leads us to ask the question: for what outcome did these Australian men and women sacrifice themselves, their ongoing health, and the wellbeing of their families?

For the hope of a 'guarantee of protection' from a great and powerful friend? A great and powerful friend whom we are now following into a potential conflict with China, under the AUKUS banner, without any public debate regarding our lack of a National

Security Strategy to justify the action taken by the previous Morrison Government? A potential conflict for which there is little to no understanding in the Australian people of the impact of such a conflict on the world and our way of life.

The authors of this chapter are not saying this as "naïve anti-war advocates." We each have decades of service in the ADF. We say this as former military officers, conscious of the cost of conflict to the men and women in the ADF, and to the broader community, and as Australian citizens concerned with the short-term, reactive, political culture in our nation.

Finally, the question of the lack of accountability. Governments often measure progress by expenditure and not results. We need to measure national and defence capabilities by results against milestones. The liberal use of the terms 'soon and sooner' by our Defence portfolio Ministers are political tools, not milestones.[48]

The Need for a National Security Strategy

Our IIER-A National Resilience Project_posed three fundamental questions: What is a resilient society? Are we resilient enough? Can we make ourselves more resilient?[49]

We postulated three key attributes of a resilient society:

- **Shared Awareness / Goals.** With shared awareness we can act rationally and prepare accordingly because we can then define a shared goal - a common aimpoint; without it, we just react to each crisis as it occurs.
- **Teaming / Collaboration.** We cannot solve our complex challenges by looking for incremental, stove-piped, quick wins; we need a team approach within our nation and, as importantly, with our neighbours and allies.
- **Preparedness / Mobilisation.** There is no verb for 'resilience'; the verb 'prepare' is the most relevant in this case. As a nation we need to prepare for future disasters / crises and not just wait to react. In addition to

> preparing, we must be able to mobilise the nation to address an emerging threat.

We fail in our approach to national security at the first step, that is, of shared awareness. Increasing levels of misinformation in both professional and social media, combined with extensive use of political spin have resulted in higher levels of complacency, confusion, and disengagement in our society.

Political divisiveness further diminishes the ability to build shared awareness and shared goals in our nation at a time where the need to do so has never been greater.

Our assumptions frame our resilience and therefore they also need to be made explicit.

Short-termism, or 'quick win' thinking, is deeply embedded in the Australian political culture; collectively we tend to focus on today and largely on our personal needs, not on future interacting and cascading risks that will impact our whole society.

Thirty years of relative prosperity in Australia, fuelled by lower trade barriers, privatisation, and deregulation, have increased our productivity and wealth, providing the resources necessary to address the challenges we confront today if we choose to act.

However, many of these challenges are themselves a result of globalisation, e.g. our extensive reliance on overseas supply chains for critical goods, leaving our nation vulnerable. Whilst the lower cost of goods has had economic and standard of living benefits, there is a very high price to ‘cheap’ in a crisis.

Australian Governments have readily accepted responsibility for national security; however, since 2013 no federal Government has had a national security and resilience strategy.

This is an unacceptable situation in today’s complex world, disrupted by the pandemic and facing growing global economic and security challenges.

The DSR 2023 is understandably focussed on our regional security concerns. However, we should also pay heed to the comments of European leaders such as Polish Prime Minister Donald Tusk, who said Europe is entering a ‘prewar’ era, cautioning that the continent is not ready and urging European countries to step up defence investment …"I don't want to scare anyone, but war is no longer a concept from the past. It's real and it started over two years ago."[50]

Conflict in Europe, the Middle East and in the Asia Pacific region seems as unthinkable today as it was in the late 1930s. But it happened, and the same warning signs are here today.

The failure to grasp the need for a comprehensive national risk assessment to underpin National Security Strategy leads our politicians to default to platform level solutions to a security crisis.

Announcing the acquisition of a small number of nuclear-powered submarines, that may not achieve full operational capability for another 20 years, will do little to address the security challenges we face today and into the near future.

Whilst this is a grim assessment there is some cause for cautious optimism. The actions we need to take are not beyond our ability to design and implement.

We have considerable expertise, will and resources in this country. In these times of uncertainty, we, the Australian people, need to act and demand more of our political system.

We should start with a comprehensive national risk assessment and a National Security Strategy, and we should hold our politicians accountable for delivering those in practicable, and not simply political or rhetorical, ways.

TWO

Shaping a Way Ahead for Australian Defence and Deterrence

This chapter is republished from my 2023 book entitled, *Australia and Indo-Pacific Defence: Anchoring a Way Ahead.* It provides a check list of key challenges which face Australia as it reworks its approach to defence. It is a shortened version of the original chapter. And provides a complement to the previous chapter.

ANY CONSIDERATION of the way ahead for Australian defence proceeds within the changing global context and in the challenges to playing an anchoring role for the democracies seeking to deter Chinese aggression in the region and beyond. It is in this context that we need to consider some of the key factors in shaping a way ahead for Australian defence.

Each of the twelve factors highlighted in this chapter have been discussed at length with Australian officials, ADF officers, or Australian strategists. And each is rooted in these conversations with regard to the growing role which Australia plays and can play in the changing geographical context where the reach of digital, information warfare, and military systems has included Australia in the first line of Indo-Pacific defence.

Crafting a National Deterrence Doctrine

Precisely because the challenges from China are comprehensive, shaping only a strategy for the ADF misses the point. Whether considering trade, information war, cyber intrusion, or political and economic engagement within the domestic society, the Chinese reshaping of the global order is not narrowly military.

And with the dynamics of change in the United States and Britain, no amount of AUKUS will solve the challenge for Australia to shape a national resilience and security strategy.

An April 2023 discussion with Dr. Andrew Carr of Australian National University focused on the challenge to Australian political culture of actually crafting a national deterrent doctrine and strategy. Carr underscored: "What are we deterring China from doing? This is not just a military task. We need to address it publicly, both to gain ongoing support from the public but also to clarify what we expect from government coordination across the whole of government to deter China.

"Deterrence is very new in the Australian experience. We have been part of a Western coalition for a very long time, but we have never had to do the kind of messaging and communication which is a crucial part of deterrence. There is not a lot of muscle memory in Australia for deterrent discourse."

China has become a different kind of competitor and adversary and partner as it changed from the reform years and building its economy to that of the China under President Xi who is combining elements of power to shape the global system more in the Chinese image.

What will Australia accept in working with its main trade partners? And what will it not? What role will foreign students from China play in Australian Australian universities? What actions by China are clearly to be countered? Which tolerated? Which ignored?

All of this is part of shaping deterrent language and narrative. What tools does Australia need to deter against which types of actions? Where does the military fit into a broader deterrent effort? Such an effort involves the broader Australia economic, social, cultural, information and security interests and not just limited to professional military competence.

Carr's key point is that such questions need to be central to Australian debate and consideration, and regularly so. There are ongoing considerations of what is to be deterred and what means need to be developed to do so.

Carr concluded our conversation by highlighting a central problem facing Western policy makers. Simply put, with the end of the Cold War and the seeming end of history and the victory of liberal democracy underwritten by the United States, policy makers saw the rules-based order as global with little clarity with regard to what are core versus peripheral interests. The term global commons came into vogue and suggested a global interdependent order in which interests were dictated by the need to deal with the gaps in the seams wherever and whenever they occurred.

Deterrence is national in character and to be effective requires clarity with regard to core interests versus peripheral interests. It also requires a realistic sense of limits. What can the nation actually do that will be seen as credible by the adversary? And will the nation have the will to do so?

As Carr put it: "The grey zone challenge comes from this global lack of clarity. With our "rules-based order" language, we tend to suggest that everything in the status quo is of interest for the West. Chinese actions in the South China Sea and Russia's actions in Crimea in 2014, called our bluff.

"Deterrence is then a policy of limits as well as focus. But it cannot remain a policy only pursued by the military, while absent from the discussions of the political class and the public."

China's relationship with Australia shaped in the past two decades cannot continue; but what kind of relationship can it be? What are its limits and what are the paths of cooperation and the focus of deterrence?

The Art of Statecraft and Crisis Management

A key capability flows from shaping an ability to craft, execute, revise, and navigate a national deterrent doctrine, namely the art of statecraft. The art of statecraft is a largely missing ingredient in Western military policy for a long time—the military in the past two decades is dispatched somewhere and comes home after largely inconclusive or negative strategic results. AUKUS certainly cannot help with this because the United States and the United Kingdom have been practitioners of the dispatch and send force practice without much correlation with strategic gain or result.

The Defence Strategic Review does underscore the importance of the art of statecraft in the way ahead for Australian defence strategy. As noted, "Statecraft must be driven and directed by a clear sense of national strategy and be coordinated across government through a clear and holistic national strategic approach. Defence's role in this whole-of-nation strategy is critical. Military power enables pursuit of a wide range of Australian interests in peacetime and is fundamental to deterring conflict, defending Australia, and denying an adversary in the event of armed conflict."[1]

But as the late Brendan Sargeant discussed with me, the Australian government faced major challenges in shaping an effective way ahead to add the kind of capabilities necessary to exercise the art of statecraft using military forces. In a meeting in 2019, we discussed this challenge at some length. As he noted at the outset of our conversation: "Globalization has unleashed one set of forces; the rise of nationalism another set of forces; and the rise of the illiberal powers yet a different set of force. Nations are trying to work out how best to protect their interests and with whom to work to do so."

This has a significant impact on the inherited alliances. There is the habitual cooperation which has underlaid the Western Alliances and that cooperation is continuing but in the context of a significant redefinition of what alliances are going to look like going forward. "Great powers like the United States are more interested in totalizing alliance arrangements than their alliance partners are likely to accept. Australians like other regional allies of the United States will seek working arrangements with a variety of regional partners to provide for our interests and work through different sorts of working arrangements to deal with our strategic challenges."

The shift is clearly from followership to engagement in working relationships where leadership is shouldered or shared differently from the great power followership role which Australia has followed first with Britain and then with the United States. Working relationships with regional or global partners around specific issues and challenges are becoming the "real" alliances. They are being built in response to specific crisis or specific problems.

For Australia, the challenge will be how to deal with global and regional crisis management. For defence, this means shaping capability which can be leveraged in a crisis and effectively used by political leadership effectively to meet the national interest. This means taking a hard look at the kind of defence force which Australia has and is developing and determining which tools are available to decision makers. It also means building a more durable and sustainable force through a crisis period.

"The ability to deploy force creates more decision space in a crisis. But you need to do that over time. That requires a robust logistical and industrial base that can give you more confidence that you can scale up during a crisis."

And the crisis management challenge requires thinking through partnerships and working relationships with allies. "When do you exercise leadership? When do you exercise followership?"

An example of how force packaging might be reworked in terms of partnerships in the region could be the working relationship between Australia and Indonesia. "We ought to be able to put together an integrated task force with Indonesia to manage a regional crisis from the low end to the high end. And a task force where either Australia or Indonesia could take the lead."

In short, according to Sargeant, "We need to think differently about our position in this part of the world and how that may drive our thinking about the capability which we need to have and to develop going forward."

Working the Geography

There are a number of different aspects of the geographical dimension of shaping a way ahead for Australian defence strategy. First, there is the question of how to leverage the geography of Australia to enhance national defence viability and survivability. Second, there is the question of how Australia can provide innovative new ways to support allies lacking geographical depth in the region. Third, there is the question of where ADF forces need to operate within the Indo-Pacific region to provide effective crisis management and deterrence effects. And, finally, there is the crucial question of how Australia works its neighbourhood effectively as an incubator against Chinese intrusions.

Here I will primarily deal briefly with the first two aspects of the geography dimension.

Leveraging Australia's geography is a key part of shaping an effective way ahead for the direct defence of Australia. Australia's geography can play a key role, in both providing for deployment and sustainment mobility as well as providing diverse launch points for longer range effects, alone or with allied engagement from Australian territory.

Again, I turned to Dr. Andrew Carr for a discussion of this aspect of shaping a way ahead for Australian defence strategy. Carr reminded me that the question of the use of territory in the direct defence of Australia has a long history. "Looking back at Australian history, there has been experience which can be drawn on as we look forward. And much of our investment in the ADF has focused on infrastructure in Australia. But with the more direct challenge posed by China, there is a re-think and re-focus going on with regard to how best to leverage Australian territory in our defence posture."

The focus now is upon "what are the key areas of Australia for defence efforts, whether population centers, bases, supply centers, production centers and so on. In effect, what is being considered is an archipelago concept in terms of understanding how the

Australian territorial chessboard can be most effectively utilized in deterrence and defence."

This has an important impact on the Australian Army, for example, as the Army shifts from a primary focus on being an expeditionary force going somewhere globally, to being a key enabler of the direct defence of Australia and leveraging Australian territory as an enabler in regional defence and deterrence. The Northern and Western parts of Australia provide significant territory in such an effort, but resources and population are scarce to do so. But new technologies—notably various autonomous technologies, such as UAVs, USVs, UUVs, and ground robotic vehicles—provide for new ways to leverage Australian territory even in the presence of limited civilian infrastructure.

If one thinks of Australian territory as a launchpad for operations into the region, then how do you organise the ADF to do so? How do you work with core allies such as the United States and Japan to share use of territory for projection of force? What kind of new basing solutions might be created to share operations between Australians and allies and to enable more robust ADF national operations?

Carr noted: "We are perhaps talking about a new alliance bargain for Australia. We would work together as coalition partners, and we would be doing tasks towards a common mission.

"But I think clarity about how Australia contributes to the alliance, what Australia is getting from that alliance are actually going to be first order questions in order to make the specific operations from Australian soil more effective.

"If we just simply have more Japanese forces and more American forces here on Australian soil, and they're replicating what the Australians are trying to do, or they're competing for use of the key locations and key airfields, and things like that, then there's going to be real challenges and impediments to operations and potentially negative public spillover from such strategic confusion."

Carr concluded: "When we think in terms of a chessboard or archipelagic metaphor, then some of the distinctions between what is specifically Australian territory and what kind of forward presence points are crucial will start to become clearer. It will be our ability to move between a whole range of access points that will be absolutely critical."

Operating Within and From Australian Territory

Re-configuring the ADF to operate across Australian geography for survivability will be a major challenge, which will take time, imagination, and money. Operating in more austere areas also raises the question of the location of manpower and the challenge of being able to move critical resources such as fuel and weapons to austere locations within Australia.

As Air Vice-Marshal Michael Kitcher, Deputy Chief Joint Operations (DCJOPS), underscored in my interview with him in April 2023: "The really good thing about Australia is the size of Australia and the amount of nothing that is in Australia. The really bad thing about Australia is the size of Australia and the amount of nothing that is in Australia."

Australia's population and economic base is in the south and east of the country; the core defence locations for projecting force into the region are in the north, north east, and north west of Australia. Northern Australia (especially the north and north west) is lightly populated without significant infrastructure and major industrial base. How does Australia have capabilities which can be used to project force into the region from Northern Australia, but the majority of the population and industrial base remains well in the south?

For example, the RAAF has a number of bare bases in northern Australia in addition to their main operating bases. But how can those bases really be used for operations in a crisis, and flexibly use all of the basing options available? How to support all these locations? How to move fuel and weapons? How to ensure the necessary level of resilience and that combat support, logistics, and health elements are available? There are no easy or cheap solutions to achieving a viable outcome.

Air Vice-Marshal Kitcher concluded: "The challenge of how we optimize the Australian geography for defence is real and is quite significant. As is how we use Australian geography for the best effect of allies and partners that we might invite to deploy here. This is an ongoing process and a real challenge."

In an additional interview in April 2023, Air Vice-Marshal Darren Goldie, the air commander of the RAAF, provided further insight into this challenge from the operator's point of view. Our engagement through two decades in the Middle East has arguably driven us down a single service route to force generation, focused on expeditionary operations, hosted from secure bases. We now need to look to evolve our approach to joint force generation from Australian territory. We don't have the level of knowledge and normative experience we need to generate regarding infrastructure across Western and Northern Australia for the Australian version of agile combat employment."

He contrasted the Australian to the PACAF approach to agility. The USAF in his view was working on how to trim down support staff for air operations and learning how to use multiple bases in the Pacific, some of which they owned and some of which they did not own. In contrast, the Australian concept he was highlighting was focused on Australian geography and how the joint force and the infrastructure which could be built—much of it mobile—could allow for dispersed air combat operations.

This meant in his view that "we need to have a clear understanding of the fail and no-fail enablers" for the kind of dispersed operations necessary to enhance the ADF's deterrent capability.

A key element of this is C2. Rather than looking to traditional CAOC battle management, the focus needs as well to be on C2 in a dispersed or disaggregate way, where the commander knows what is available to them in an area of operations and aggregate those forces into an integrated combat element operating as a distributed entity.

Goldie commented: "We are developing concepts about how we will do command and control on a more geographic basis. This builds on our history with Darwin and Tindal to a certain extent, although technology has widened that scale to be a truly continental distributed control concept. We already are familiar with how an air asset like the Wedgetail can take over the C2 of an air battle when communications are cut to the CAOC, but we don't have a great understanding of how that works from a geographic basing perspective.

"What authorities to move aircraft, people and other assets are vested in local area Commanders that would be resilient to degradation in communications from the theatre commander—or JFACC? We need to focus on how we can design our force to manoeuvre effectively using our own territory as the chessboard."

Air Vice-Marshal Goldie underscored that the ability to work with limited resources to generate air combat capability is exercised regularly by the normal activity of 75 Squadron, flying F-35s in Australia's Air Combat Group. This squadron operates from RAAF Base Tindal in the Northern Territory and as Goldie put it: "they have to operate with what they have in a very austere area."

By learning how to use Australian territory to support agile air operations, and to take those capabilities to partner or allied operational areas, Australia will significantly enhance its deterrent capabilities going forward.

Force Distribution and Integration

In my book with Ed Timperlake on the building of the kill-web ecosystem for concepts of operations and force development, this is how we described force distribution and integration in shaping a way ahead for force development: "Force packages or combat clusters are deployed under mission command with enough organic C2 and ISR to monitor their situations and integrate the platforms that are part of that combat cluster and to operate effectively at a point of interest. Within that combat cluster, the C2 and ISR systems allow for reachback to non-organic combat assets which are then conjoined operational for a period of time to that combat cluster and becomes part of an expanded modular task force.

"Such an approach and capabilities are the essence of what a kill-web enabled force is and how such integratability can close the geographical and combat seams which twenty-first-century authoritarian powers are focused on generating. This allows for the kind of escalation management and control crucial for the competition with the twenty-first-century authoritarian powers.

"It is not about getting to World War III as rapidly as possible or generating nuclear exchanges early in a widening conflict. It is about escalation control and management, and an ability to close seams which adversaries seek to open to gain significant escalation dominance as they expand the reach and range of those twenty-first-century authoritarian powers.

"A shift to a kill web approach with regard to force development, training and operations is a foundation from which the U.S. and its allies can best leverage the force we have and the upgrade paths to follow. For this approach to work, there is a clear need for a different kind of C2 and ISR infrastructure to enable the shift in concepts of operations . . .

"A very different kind of C2 and ISR infrastructure is at hand to build enablement for distributed operations. The new C2 and ISR infrastructure requires rethinking considerably the nature of decision making and the viability of the classic notion of the Observe, Orient, Decide and Act (OODA) loop. If the machines are fusing data or doing the OO function, then the DA part of the equation becomes transformed, notably if done in terms of decision making at the tactical edge. The decisions at the edge will drive a reshaping of the information about the battlespace because actors at the tactical edge are recreating the information environment itself."[2]

The United States and the ADF are on convergent paths with regard to such developments, sharing technology, C2/ISR arrangements, and a number of key elements to shape a distributed force operationally with integrated combat effects. The level of collaboration between the United States and the ADF is historically unprecedented and opens up the possibility of scalability in terms of the operations of the two forces. Although the two allied militaries closely collaborate, they represent different nations with overlapping but different interests. Force distribution will focus on the survival of the national forces but with an ability to work with the other nation in providing for strategic depth for the other.

For the ADF, force distribution starts with leveraging the national territory and being able to have a resilient force supported from national territory. It proceeds out into the neighbourhood with the ability of the combined air and naval force to project power out to Australia's first island chain with the Army able to move rapidly forces to reinforce partners in the region when facing pressures in the region.

This gets back to the geography factor mentioned earlier, namely where ADF forces need to operate within the Indo-Pacific region to provide effective crisis management and deterrence effects. The ADF is a modest force, well trained in coalition operations, but the direct defence of Australia even with expanded reach means that the ADF's core capability needs to be able to operate from the Australian "sanctuary" outwards to the first island chain, and when necessary for deterrent effect or crisis management, to operate beyond the first island chain.

Where does the ADF primarily need to operate? With what kind of joint force? With which allies and partners? And with what capabilities?

The shift from the wars of choice in the Middle East to the defence of Australia within its region is a major one which clearly will affect force structure development. As Air Vice-Marshal Michael Kitcher, Deputy Chief Joint Operations (DCJOPS), put it in my April 2023 interview with him: "The focus in this period, up to say 2017, for CJOPS was on operations in the Middle East whilst managing operations in our region. We clearly have leveraged the earlier experiences in our renewed focus on the conduct of Operations, Actions and Activities (OAA) in the Indo-Pacific. We are focused on developing a theatre campaign plan to translate strategic guidance into the OAA we execute in our region to achieve our desired objectives.

"We are focused on ways we can operate as a joint force to optimise our regional OAA to have the maximum positive effect in supporting our theater campaign plan. You don't get the maximum benefits from a joint force unless firstly the services provide you with trained personnel capable of executing joint missions and then HQJOC, through focused joint planning, maximises the potential of the individual components. We have made good progress along this path but still have a way to go."

Air Vice-Marshal Kitcher highlighted that we are "now squarely focused on managing operations in a coordinated fashion in our region." And this means both, how to get the best joint force effect but also how to coordinate the ADF effort with core allies in also getting the best proper coalition effect.

Obviously in working with coalition partners, national sovereignty has to be respected but at the same time for effectiveness in operations, coalition forces need to operate in an integrated manner. This is a key tension which needs to be managed; notably in crises where the government of the day will make decisions about the allowable operations of their national forces, these individual decisions may challenge the effectiveness of a coalition force.

This is a challenge which CJOPS has to be prepared to deal with in both exercises and real-world operations. Kitcher underscored: "Planning and exercises prepare the way for joint and coalition capabilities but executing them in an actual operational situation requires agility and flexibility of command by CJOPS and his staff, and our parallel staff in the various coalition headquarters."

Air Vice-Marshal Kitcher emphasised that working with partners to deal with challenges in the region has clearly grown in importance for both deterrent and operational impacts. The relationship with U.S. forces has certainly become closer. He mentioned an upcoming CPX exercise with the U.S. Indo-PACOM command in which the ADF and the United States will run a detailed CPX on a regional scenario together. The cooperation with both Japan and India is also growing.

And with Australia's regional defence emphasis, joint operations will need to focus on regional partners in the Australian neighbourhood. This will see more emphasis on building regional expertise and continued engagement with regional countries through relationship building, languages, cultural awareness, and local knowledge. This can provide an important aspect of Australian leadership in a regional military coalition but dependent on the crisis, a differentiator for Australian involvement as well.

As Air Vice-Marshal Kitcher summarised their job: "We've got a responsibility to make sure that we optimize how well the joint force works together for the greatest positive effect and present the best possible options to government on how that force might be employed. We've also got a remit to ensure that we can work as closely and as efficiently and effectively as possible with our regional partners in both peacetime HADR situations and potential crisis situations.

"We've got a responsibility to be as efficient and as effective together among like-minded nations' militaries. If we are not careful, uncoordinated actions in our region will overwhelm smaller countries and not have a positive effect. Planning and conducting OAA together ensures we present a much more credible regional security capability than we do as individual nations. "The militaries have a large part of the responsibility to generate how we can do so. And then it's up to individual governments to determine how those forces will be employed at any one time or in any one set of security circumstances."

Force Distribution, Sustainment, and Logistical Support

To ensure enhanced survivability, the ADF is looking to distribute over Australian territory more effectively. But this makes logistical support for distributed forces a major strategic challenge. And with the changing threat calculus, the force needs to have greater endurance which requires enhanced sustainability.

So the question is, how best to combine distribution of combat forces, with effective logistical support, but have credible sources of supply that can sustain the force?

In many ways, this poses a significant strategic triangle — force mobility, sustainability, and logistics — which has to be built and operated in the period ahead to have an effective ADF and, of course, the ADF is not alone in terms of meeting this challenge. Most notably, its major warfighting ally in the Pacific, the United States, faces the tyranny of the Pacific in dealing with this strategic matrix.

I have discussed this challenge with Colonel David Beaumont of the Australian Army several times in the past few years as he has worked on logistics issues his entire service life. Currently, he is the director of Joint Professional Military Education at the Australian Defence College located in Canberra.

As Beaumont characterised the challenge in our latest discussion in April 2023: "Sustainment and logistics capabilities determine the endurance of your force. They shape the ability of your force to remain operable. They determine how your force can sequence its operations and operate at the tip of the spear. It can be described as the arbiter of opportunity to paraphrase Thomas Kane. By that I mean, it determines when the force can and cannot act."

We are experiencing a major shift from just-in-time wars and just-in-time delivery systems to facing the challenge of response to crises created by adversaries which will challenge our ability to act, endure, and prevail.

As Beaumont put it: "We have been used to certain ways of operating in the past 20 years or so. We have operated in surges and cycles that have been well planned in advance and shaped our routines. Forces have been allocated on the basis of what we can reasonably sustain. For a country like Australia, we have been able to choose judiciously the forces we can operate operate with because we know we can sustain them at the right moment and with the right resources."

The challenge now is to prepare for a different scale and intensity of conflict which simply does not comply with limited sustainability and just-in-time logistics. Beaumont added: "We will need now to operate at the maximum of our potential and that requires logistics resources and sustainability planning to suit."

We turned to the real challenge of getting procurement systems in Australia or the United States to be able to prioritise sustainment and logistics as a strategic issue rather than a residual one.

Beaumont argued that Western military acquisition systems have for a long time prioritised platform acquisition over operational sustainability and preparedness, with corporate success often defined by the perceived effectiveness of platform delivery programs. This emphasis means that moneys tend to be drawn from sustainment or logistics budgets to pay for new platforms or cost over runs of platform programs.

How then does one change this culture and focus?

Beaumont noted that one way to do so that is being started in Australia is to deal with specific commodities capability needs to be dealt with as a program in its own right, such as the newly launched guided weapons program. It is also important to go beyond headline logistics deficiencies and resolving broader sustainment gaps across the force.

When one considers the problem of mobilisation of resources from the general economy, it is easier to conceptualise rather than do or fund. If Australia wishes to pursue greater self-reliance in stocks, then mobilisation is an inevitable subject which needs to become real in terms of programs and funding.

There is the question of enhancing production with allies in order to have an allied-wide approach to production in a new arsenal of democracy model. But the challenge remains for each of the countries involved in joint production or acquisition of stocks available in times of crisis to the national forces.

Then there is the question of logistical means to move stocks to forces which themselves are working the art of force mobility. The earlier discussed interview with the Air Commander Australia highlighted his concern with an enhanced ability for air mobility from diverse locations in Australia. But how to move the parts and supplies necessary to support such an agile operating RAAF?

And with a large territory, how will Australia produce, stockpile, and move the supplies necessary for itself and the allies who are using the Australian territory? This requires an effective national and allied interoperable IT system for logistics enterprise management as well as the ability to use maritime, rail, or road systems to move supplies to the point of need.

Force mobility, sustainability, and logistics have become a strategic triangle shaping the capability for force endurance, effectiveness, and relevance to conflict with the authoritarian powers in the twenty-first century.

Acquisition and Force Development

Along with the strategic shift towards distributed forces and operations prioritised in their neighbourhood, the build out of the ADF in the years ahead needs to be part of the great software evolution underway, and the expanded role of autonomous systems for the operating forces.

One example of the change is occurring in the domain of maritime autonomous systems. The nature of this change was highlighted in an interview which I did in April 2023 which followed up on the 2022 interview cited earlier with Commodore Darron Kavanagh, Director General Warfare Innovation, Royal Australian Navy Headquarters.

Maritime autonomous systems don't fit into the classic platform development mode or the sharp distinction between how particular platforms operate or perform and the various payloads they can carry. They are defined by the controlling software and the payloads they can deliver individually or as a wolfpack with the role of platforms subordinated to the effects they can deliver through their payloads. The software enables the payloads to be leveraged either individually, though more likely in combination as a wolfpack or a contributor to a combat cluster.

We started our discussion by focusing on mission threads as a way to understand the role and contribution of maritime autonomous systems. What missions does a combat commander need to accomplish? And how can maritime autonomous systems contribute to a mission thread for that combat commander, within the context of combat clusters?

As Commodore Kavanagh underscored: "One of the issues about how we've been looking at these systems is that we think in terms of using traditional approaches of capability realization with them. We are not creating a defence capability from scratch. These things exist, already, to a degree out in the commercial world, regardless of what defence does. AI built into robotic and autonomous systems are in the real world regardless of what the defence entities think or do. And we have shown through various autonomous warrior exercises, that we can already make important contributions to mission threads which combat commander's need to build out now and even more so going forward."

And that is really the next point. The use of maritime autonomous systems is driven by evolving concepts of operations and the mission threads within those evolving CONOPS rather than by a platform-centric traditional model of acquisition. Commodore Kavanagh pointed out that traditional acquisition is primarily focused on platform replacement and has difficulty in supporting evolving concepts of operations.

This is how he put it: "We're good at replacing platforms. That doesn't actually require a detailed CONOPS when we are just replacing something. But we now need to examine on a regular basis what other options do we have? How could we do a mission in a different way which would require a different profile completely?"

Put another way, combatant commanders can conduct mission rehearsals with their forces and can identify gaps to be closed. But the traditional acquisition approach is not optimised for closing such gaps at speed through the use of disruptive technologies. The deployment and development of autonomous systems are part of the response to the question of how gaps can be closed or narrowed narrowed rapidly and without expensive solution sets.

In an interview I did earlier this year with a senior U.S. Navy commander, he identified the "gaps" problem. "Rehearsal of operations sheds light on our gaps. if you are rehearsing, you are writing mission orders down to the trigger puller, and the trigger puller will get these orders and go, I don't know what you want me to do. Where do you want me to be? Who am I supposed to check in with? What do you want me to kill when I get there? What are my left and right limits? Do I have target engagement authority?

"This then allows a better process of writing effective mission orders so that we're actually telling the joint force what we want them to do and who's got the lead at a specific operational point. By such an approach, we are learning. We're driving requirements from the people who are actually out there trying to execute the mission, as opposed to the war gamers who were sitting on the staff trying to figure out what the trigger pullers should do."

But how to close the gaps?

As Commodore Kavanagh argued: "We need to deliver lethality at the speed of relevance. But if I go after the conventional solution, and I'm just replacing something, that's actually not a good use of my very finite resources. We need to be answering the operational commander's request to fill a gap in capability, even if it is a 30% solution compared to no solution on offer from the traditional acquisition process."

These are not technologies looked at in terms of a traditional acquisition process which requires them to go through a long period of development to form a platform which can be procured with a long-life use expectancy. Commodore Kavanagh simply pointed out that maritime autonomous systems are NOT technologies to be understood in this manner.

"We build our platforms in a classical waterfall approach where you design, develop and build a platform over twenty years to make them excellent. But their ability to adapt quickly is very limited. This is where software intensive systems such as maritime autonomous systems are a useful complement to the conventional platforms. Maritime autonomous systems are built around software first approaches, and we are able to do rapid readjustments of the code in a combat situation."

And the legacy acquisition approach is not well aligned with the evolution of warfare. Not only is the focus changing to what distributed combat clusters can combine to do in terms of combat effects, but the payload impacts at a point of relevance is also becoming of increased salience to warfighting approaches.

What is emerging clearly is a need to adapt more rapidly than what traditional platforms and their upgrade processes can do. Gaps will emerge and need to be closed not just in mission rehearsals

rehearsals but in the combat operations to be anticipated in the current and future combat situations.

And to endure in conflict, it will be crucial as well to protect one's core combat capital capabilities and platforms which calls for increased reliance on capabilities like maritime autonomous systems to take the brunt of attrition in combat situations as capital ships become mother ships rather than simply being the core assets doing the brunt of combat with whatever organic capabilities they have onboard.

As Commodore Kavanagh noted: "The nuclear-powered submarine is absolutely necessary for what we need to do for our defence in depth, but what we're focused on with maritime autonomous systems completely complements it, because what I want to do is ensure that the dangerous stuff gets done by the autonomous forces as much as possible, because we can rebuild that capability much more rapidly. We can actually restore it whereas we can't restore a nuclear-powered submarine quickly if lost."

Commodore Kavanagh closed our discussion by emphasising the crucial need for Australia to have an ability to stay in the fight in case of conflict in the Pacific. He argued that having their own abilities to innovate in autonomous systems areas was part of such a desired capability.

"Resilience in a combat situation is an ability to be able to experiment and adjust on the fly. To have an enduring force that can operate until statecraft can shape an end state, the warriors and their support community must adjust the combat force rapidly to the real-world combat conditions. By shaping a deployment and ongoing development process in the maritime autonomous systems area, we are contributing to such a combat capability."

Another take on the change in acquisition and the dynamics of change was highlighted at the Williams Foundation seminar held on 28 September 2022. At that seminar, the head of Force Design in the Australian Department of Defence, Major General Anthony Rawlins, addressed directly the question of how the ADF could realistically and effectively ramp up its capabilities in the midterm.

This is how he put it: "Has the hardening of expensive,

exquisite, arguably irreplaceable platforms now reached its logical zenith? This is manifest in the arguments for the cheap or the expendable as a supplement or potentially a replacement for expensive crewed platforms going forward.

"Defence is not just investing in exponential developments in autonomy, artificial intelligence, remote sensing, etc. as an R and D line of effort. But defence is doing so with a view to fielding capability in the immediate short term. And it hardly meets the definition of survivability to be investing in platforms and capabilities that are designed to be expendable…"

The Arsenal of Democracy: The Role of Australian Defence Industry

For Australia to have an enduring force and resilience through a crisis requires having supplies on hand or that which can be generated at home during that crisis. For this to happen, two key changes will have to occur.

The first is for the Australian defence industry to be expanded and the kind of ecosystem put in place within the economy and society to endure through a crisis period. This will take manpower, technology, manufacturing capability, and money.

The second is for the democratic partners of Australia to work much more closely together to shape an interactive arsenal of democracy. The blunt fact is that Australia cannot act as if the United States is the arsenal of democracy. The United States has reduced its defence industrial base dramatically over the years, as well as its industrial base. Australia is simply not an industrial country. One cannot assume that mobilisation of supply will be a simple switch-turning exercise. In today's world, it has to be built and funded. This is not an easy task nor one that is politically popular as well.

The Australian Department of Defence has favoured Foreign Military Sales arrangements with the United States. This has provided the ADF with proven products at a predictable cost but with overseas suppliers as essential providers of the just-in-time parts.

But a crisis in the Pacific will make such a strategy as one which will make the ADF a one-month duration military. The need is for a mix of Australian-supplied and stockpiled parts to be available to allow the ADF to operate at the level needed during a crisis.

The shortfalls evident during the current Ukraine War make it clear that simply depending on the United States as a supplier is inadequate even without the problems of transport in a crisis or the U.S. military getting priority in a crisis.

Frankly, the most credible way ahead is for the United States, Australia, and close allies to shape an acquisition strategy which prioritises standardisation of items such as munitions and build them across a global geopolitical industrial base. This will go against the grain of how the United States drew down its defence industry with the famous "last supper" and reduced the size of the contractor pool. It also goes against the legacy of the past twenty years of those prime contractors outsourcing key parts and subsystems to a national or, many times, global supply chain.

We are talking about major change in the United States, Australia, Europe, and Japan to recognise that having surge production capability is more important than simply narrow control of markets. The 20 May 2025, agreement between President Biden and Prime Minister Albanese is clearly a step in the right direction.

An April 2023 discussion with Dr. Alan Dupont provided some important insight on how to shape a way ahead with regard to Australian defence industry and the challenge of shaping an allied arsenal of democracy. We argued that discussing sustainability in terms of a defence industrial base is too narrow an approach.

It really is about shaping the entire eco-system for sustainable defence forces, which includes specific defence companies, new acquisition approaches, companies that support the core capabilities which defence taps into but are not specifically defence companies per se and tapping into new logistical and support approaches to support distributed force.

As Dupont underscored: "I think we should move away from

this defence industrial base language which can be very clunky and twentieth century. People think in terms of big factories and production and development cycles of 20 years. We need a very different focus."

Dupont laid out a methodology for building what he considers to be an appropriate Australian defence industrial effort. As it stands now, Australia is almost entirely dependent on overseas supplies and when Australia orders what it needs, it joins the queue along with other customers, with no certainty to be supplied in a timely manner. Add to this the tyranny of distance facing the transportation of military parts to Australia, and you have a perfect storm facing the Australian defence in terms of conflict.

To deal with this challenge, Australia needs to enhance its sovereign defence production capabilities. But to do so, Dupont suggests the need for a realistic methodology to shape the way ahead.

What does Australia need in terms of defense capabilities over the next two decades? How much of what it needs could realistically be produced in Australia? What can it do with co-development or co-production with key allies? And what will it simply have to procure from allied countries and producers?

In those areas where it feasible to build sovereign capabilities, a new development approach is needed. Many of the dynamic new capabilities being used by defense forces come from smaller more innovative firms. Australia has such firms but there is no Australian government policy to support them or to ensure that they have the capital to grow. There is a need for an Australian industrial policy in this area.

In areas where Australia could produce for its own needs, the government should commit to a South Korean, Israeli, or Swedish path of growing for exports. He pointed out that South Korea now exports 17 billons of dollars of exports which provides a key pillar for its own defence.

In addition, to discussing his methodology for the development of Australian sovereign defense industrial capabilities, we discussed the strategic direction of defense and how best to support it.

Defence forces in the Pacific for the liberal democracies are focusing on force distribution for survivability.

There are new technologies to support force distribution such as synthetic fuel production and 3D printing in the field. New approaches to sustaining distributed forces through a relevant development and production support are crucial to provide enhanced capabilities for distributed forces.

New platform/payload combinations are being introduced through such sectors as aerial and maritime autonomous systems. How will Australia support this effort? How will it do so in a way that allows for exportability? How will it work with core allies to enhance the rapidity of change in this area?

Cost effective and expendable platforms carrying a variety of payloads are a key element of the new defense equipment ecosystem. How will this ecosystem be supported and thrive? Most likely not with old acquisition approaches and older concepts of a "defense industrial base."

A reworking of the Australian approach to supplying its forces is required. But it should be done a realistic manner but with a focus on the force structure changes taking place and the need to help sustain a distributed defense force both now and in the future.

"Impactful Projection"

The DSR argues for what it calls "impactful projection" for the ADF. This is how the DSR characterizes the requirement: "The defence of Australia lies in the collective security of the Indo-Pacific. The defence of Australia's national interests lies in the protection of our economic connection with the world and the maintenance of the global rules-based order.

"Accordingly, the Australian Defence Force (ADF) must have the capacity to:

- defend Australia and our immediate region;
- deter through denial any adversary's attempt to project

power against Australia through our northern approaches;
- protect Australia's economic connection to our region and the world;
- contribute with our partners to the collective security of the Indo-Pacific; and
- contribute with our partners to the maintenance of the global rules-based order.

"As most of these objectives lie well beyond our borders, the ADF must have the capacity to engage in impactful projection across the full spectrum of proportionate response. The ADF must be able to hold an adversary at risk further from our shores."

But this gets again at the geography and effect problem. Which geography can the ADF operate with enough impact to make a difference against the Chinese?

Personally, I think this breaks down into two different answers.

The first is what the ADF can do operating with the United States or Japanese forces or both deep in the Pacific from Australia's point of view but closer to Chinese military actions in either their first or second island chain? And secondly, how capable the ADF is to operate by building defence in depth within Australia and then operating up to and including its first island chain?

The key tissue for doing either operational matrix is the capability of the C2/ISR networks to support ADF operations both nationally and in coalition. The longer the range the harder the challenge and the more likely to face significant threats to the viability of those networks.

And the longer the distance and closer to China the absolute imperative for Australia and American targeting to be integrated, for the simple fact that conventional long-range strikes are inherently related to the questions of the run up to nuclear deterrence.

There is no hard break between targeting and adversaries C2/ISR systems for longer range strike conventionally and the spill over into nuclear use issues.

Paul Bracken has put this challenge rather clearly: In an article published in *The Hill*, Bracken underscored the threat of inadvertent use of new technologies for conventional modernization upon escalation management.

Failure to think about new military technologies makes it more likely that these systems could be used in reckless ways and, if used, could lead to unplanned escalation. It is important to take this problem seriously, because we are in the early stage of a long- term arms race with advanced technologies such as artificial intelligence (AI), hypersonic missiles, cyber weapons, drones and the like….

Drones, cyber, AI, hypersonic missiles, anti-satellite attack and other advanced technologies are central to U.S long-term competition with China and Russia. Some of these technologies have spread to North Korea and Iran. These two nations, for example, already are major threats in cyber war, and both operate armed drones.

Here we see the new escalation problem facing the United States. Using these technologies against terrorists and insurgents, or to disrupt a weak power such as Iran's uranium enrichment, the chances of a large eruption in violence by an enemy's response are low. Terrorists and insurgents lack the weapons to strike back at the United States in a meaningful way.

But used against China, Russia or North Korea, the risk is altogether different. Even in a limited war with conventional weapons the new technologies could become highly destabilizing. More, these are nuclear-weapon states. In peacetime, China, Russia and North Korea do a good job of controlling their nuclear forces. There are no reports of accidents or unintended missile launches by any of them. They are cautious and disciplined when it comes to these weapons.

But this is in peacetime — caution and discipline are easy because there are no stresses on their leadership or military command systems.

Crises and limited wars aren't like that. In an intense crisis or limited war, all bets are off — or, at least, all bets need to be recalculated. And this is the point: Any policy review needs to assess the escalation potential of their use. Basing reviews on the way they are used against terrorists and insurgents overlooks the critical difference that heavily armed nations bring to the problem.

They can strike back at the United States or its allies, and this ability calls for a different type of review that goes beyond collateral damage. It must focus on the likelihood of counter-escalation.

To control escalation, the United States must take account of the fog of war not only in our own forces, but in the enemy forces as well. Against a nuclear weapon state, the United States isn't simply taking out targets; it is playing a larger game of mutual risk- taking. Any review that ignores this point misses the essence of the decision facing U.S. leaders."[3]

The DSR highlights two new acquisitions which are designed to give Australia a seat at the table in making decisions about "impactful projection" involving long-range strikes, namely the acquisition of SSNs and of Tomahawk missiles.

Australia is acquiring up to 220 Tomahawk Land Attack Missiles and will be the first country outside of the US and the UK to acquire them and these weapons will be integrated into the Royal Australian Navy's Hobart-class destroyers.

The other part of "impactful projection" from my point of view even more significant in dealing with the Chinese challenge is to provide for Australia's direct defence and enhanced capabilities to operate from its own territory and up to and including the Australian first island chain.

As Brian Weston suggested in a 2020 series of articles on the Williams Foundation website but were originally published by the Australian Defence Business Review: "the notion of Australia's First Island Chain brings a clearer conceptual basis for force development and operational planning, a lesser dependence on the complexities and national interests of partners and allies and yet, the region remains of critical relevance to Australia's security. So, is Australia capitalising on these realities by devoting enough effort to the detail of how Australia can defend and dominate the nation's "Red Zone"?"[4]

Doing so is very challenging to the current ADF but several of the pieces of the DSR if implemented could well allow Australia, for the ADF by itself is not enough to be able to do so, to operate from its own territory, notably the North and build a C2/ISR and force mobility force able to dominate the nation's red zone.

This has the other advantage of providing a mobile sanctuary for the core allies of Australia to reinforce their ability to fight in a crisis in terms of sustainment and support, as well as working effective coalition operations with partners in the region.

But doing both – a significant commitment to long-range strike and an ability to build out defence of the nation's Red Zone is frankly far beyond the current budget commitments of the current government. There is also the really critical question of the C2/ISR capabilities to operate a force at distance versus one leveraging one's continent projecting out into the red zone.

The challenge of such a build out can be seen in the case of one program which can provide for both Red Zone operations and support longer range strike, but will not be enough by itself, namely, the Triton program.

Wing Commander Keirin Joyce, Program Chief Engineer RPAS (MQ-4C Triton) for the RAAF highlighted in an interview in April 2023 that with the U.S. Navy and the RAAF both operating the Triton, working cooperative operations can clearly be envisaged as Australia and the U.S. Navy will compliment areas of operations of significance to both countries to enhance the ISR/C2 capabilities of both. And as the ADF builds out its longer-range strike capabilities, having the Triton as an asset to assist in the targeting process will be important as well.

Triton comes at a key time in the evolution of ADF capabilities to enable longer-range effects from Australia out into the region. But Joyce commented that what will be interesting to note 'is this enough'? He thinks Australia will need even more assets, and uncrewed/automated/autonomous assets are probably the answer in the current challenging climate of attracting and retaining workforce."

If one gets specific and concrete about what the ADF has to operate with and where it must operate, I think one needs to think in terms of impacts on the Chinese militarily and not just long-range strike. Without doubt it is important to have national long-range strike precisely to give Australia a seat at the table in the case of broader conflict in the Pacific.

But this reminds more of the years of work I have done with the French on the question of influencing the Americans to do the right thing from their point of view in European defence, and having their independent nuclear deterrent. Having independent long-range strike is the currently conceived functional equivalent.

Working in the Neighbourhood

The ability to work effectively in the neighbourhood from a military point of view starts with building a more resilient Australia within which the military can operate and be sustained as a sanctuary from which to project forces in the region. This is a major task which if pursued as a priority will take time, effort, innovation, and investments.

Technology will be part of such an effort. Building secure telecommunications channels that can be used in a crisis, the use of new and innovative ways such as using UAVs for delivery systems to bare bases and locations, and the use of USVs and UUVs to create ISR grids which can feed into diversified C2 networks for the command of distributed forces are examples illustrating what a comprehensive effort would look like.

The build out of locations from which repairs can be performed for military equipment, including ships and airplanes, will require innovation in terms of new ways such as 3D printing to allow for a more diversified set of locations from which to do so for both the ADF and allied militaries.

The Maritime Border Command works closely with partners in the neighbourhood to work the many issues involved in shared situational awareness and joint operations to provide for security to Australia from the sea.[5]

This could be expanded into new information sharing with the neighbours in terms of basic security and maritime awareness through shared acquisition of USVs and sharing of information generated from these USVs. Currently, the Australian government is using an Australian company to build USVs generating ISR information useful for such tasks.[6]

Why not build out shared builds, shared operations, and shared data to work an ISR grid for security awareness across the region?

To provide for the maritime and air security of Australia, a build-up of ISR capability and associated C2 is underway. Why not expand this capability by using the new generation of USVs, UUVs, and UAVs which are being tested, developed, and built? By building a dense network of surveillance to support the Australian sanctuary, there is the added benefit of doing so to aid allies in terms of crisis as well to provide support for rotational allied forces.

Such an approach has important effects on the build out of the services and the joint force. The RAAF remains the tip of the spear. There is no element of military power more capable of signalling to an adversary your early involvement in a crisis or shaping a deterrent effect than airpower. The question will be how to bulk up the deterrent effect from the manned element of the force and at greater distance. But airpower has the flexibility of providing a rapid scaled effect.

The resilience challenge comes into play here. How will Australia have adequate strike capabilities? How to build or store adequate munitions for the force? The answer in part is using Australia's unique test ranges in a broader alliance effort to build out a new generation of missiles which are built jointly with the United States and like-minded states. The example of what Norway has done with the Naval Strike missile is one to be considered by Australia.

The Australians have stood up a weapons enterprise, but it will need to do much more than being a conduit for importing extant American missiles. It needs to become a fulcrum for generating change in the weapons side of the equation for Australia and by so doing for the broader set of alliance partners of Australia.

The Royal Australian Navy faces a major challenge in terms of sorting out the mix and match platforms within the operating force.

But the way ahead in my view for the maritime force is to operate within a kill-web ecosystem where ships can both contribute to and benefit from third-party targeting to develop distributed but integrated firing solutions.

This was underscored in an interview I did with the U.S. Navy sub commander in an interview conducted in Honolulu in April

2023. Rear Admiral Jablon underscored the nature of the shift as follows:

The submarine force is now becoming part of the 'combat clusters' that you're talking about instead of an independent operator. In the Cold War, we operated independently, alone, and unafraid. During the land wars, we started becoming part of the joint force as we provided land fires via the TLAM.

Now, we are fully integrated with the joint force in terms of targeting and communications. But, of course, we can also conduct independent operations as the 'silent service' when directed.

But what capital ships can Australia build and maintain effectively? What class of ships? What will be the interaction between these capital ships and autonomous systems?

There is a whole world of innovation in the maritime domain underway which Australia must tap into because of its limited manpower and manufacturing resources.

Rather than taking a platform replacement perspective or build strategy, the focus needs to shift to a distributed but integratable kill-web ecosystem within which ships can operate. This is where innovation can be driven in the rapidly evolving robotic world.

Rather than seeing autonomous systems in the short- or medium-term creating ghost fleets, their role will be to expand the range, capability, and lethality of capital assets.

Rather than looking simply at the organic capability on a specific platform, we will consider surface ships using such capabilities as becoming mother ships and submarines will share in this development as well.

But then how to build ships in a world altered by the arrival of USVs, UUVs, and UAVs?

And how will the rebuild of the fleet interact with the strategic shift in terms of the Australian Army?

The Australian Army given its operation of the land is the key force in terms of working in the neighbourhood. Re-thinking Army's role which has been called for in the DSR is closely connected with defining what the defence perimeter for Australian direct defence is and how best for the ADF to operate in that space.

In a 2022 interview which John Blackburn and I had with John

Blaxland, we discussed this issue and its impact on the way ahead for the ADF. We focused on what we agreed most logically defined the defence perimeter: operations from the continent to Australia's first island chain, which is outward from the continent to the Solomon Islands and across to Papua New Guinea, Timor L'este, and Indonesia.

This is how Blaxland defined how to shape an approach for this strategic space with the ADF operating an appropriate manoeuvre force supported by appropriate infrastructure in Northern and Western Australia which could support such a force.

We need to be mindful of our history and our geography and our neighborhood is a neighborhood with a history of violence. When we faced existential challenges in the past, we have had to forward deploy into the island chain to Australia's north.

He argued that the initial efforts to project such a force did not work out all that well in the early years of World War II but by 1944, Australia had sorted out a more effective capability to operate outward into the first island chain.

He noted:

When we faced an existential crisis in 1942, it came through the archipelago, through Indonesia, through Papua New Guinea and the Solomon Islands. That is increasingly contested space today, and it's all the more important that we invest in a strategy that capitalizes on the relationships with those countries in the archipelago.

From a force structure point of view, that means actually thinking about how we structure our forces to enable deployment in that space and to operate alongside the neighbours, but in a contested environment.

As a former Army officer, he discussed how he saw the future of the Army working with the joint force in the strategic space defined by operating from the continent to the first island chain. He argued that defence diplomacy coupled with an enhanced ability of the Army to operate forward in support of the air and sea forces was critical.

To project airpower at range likely would require maintenance of lily pad–like forward operating bases for which a ground force has an important defensive role—including capabilities that would ensure overmatch against a would-be adversary—including mobile, protected armoured platforms with the ability to reach out and touch someone at range.

With regard to defence diplomacy, this is what he underscored:

We have for the last two decades basically been distracted by operations in the Middle East, what I call our niche wars in Afghanistan and Iraq. We've dropped the ball in terms of building relationships, understanding the culture, the language, the people, the networks in our neighborhood.

We are starting to reinvest in that space, but we've got a long way to go. Very few of our seniors and our middle level commanders and managers speak the relevant languages, for example.

We have designed a force to plug and play with the Americans, but that is not what we need for the future. Interoperability with the Americans remains important for sure, but Australia needs to be able to conduct its own operations in the neighborhood in a self-reliant manner, as well as alongside neighbours.

What this also means is that Northern and Western Australia need to see significant infrastructure development to sustain and operate such a force. And frankly, this is a whole of government issue, and not just about what the Department of Defence can fund.

Such a shift towards enhanced capabilities for direct defence of Australia clearly has implications for core allies like Japan and the United States.

Blaxland argued:

The best thing we can do is make Australia more self-reliant, more resilient, more able to defend its own turf, and collaborate with neighbors to defend our common strategic space.

If Australia provides a firm base, and if we have a leavening effect on our own neighborhood, then that takes away the stress of planning for others who might be thinking about other contingencies, and a key point to my mind is this is all about deterring war, deterring the prospects of war. We need to dissuade adventures from competitors and would-be adversaries while being reared to fight and win when deterrence fails.

All of this leave a significant impact on the way ahead for the Australian Army. It plays a key role within Australia but what is the government asking it to do going forward? And what is it willing to invest in so that it could do so?

A key implication of the shift to the neighbourhood is finding ways for the Army to move from Australian territory to areas within the neighbourhood where it might need to go.

With a significant shortfall on lift—sea or air—how will this happen?

What kind of RAN will be built to support this?

How will Army aviation be able to move quickly if it continues to not have access to tiltrotor which has revolutionised the ability of the USMC to move ground forces rapidly across the Pacific chessboard?

In my interview with the chief of Army in April 2023, Lieutenant General Stuart Simon provided some of the answers which he thought the Army could provide for the strategic shift. Lieutenant General Stuart spoke at some length to the territorial presence role of the Australian Army as underwriting the ability for direct defence and shaping an effective foundation for joint force power projection from Australia.

Lt General Stuart underscored:

At the foundation and during my presentation, we emphasized the importance of being able to leverage Australian geography for strategic purposes. The Army is located in 157 locations around the country, from the most northern tip of Cape York to Tasmania and from the west to the east coast. And our connection into local towns, cities and communities is through these 157 locations where our people are located.

One of the key design principles for our Army Objective Force is what we call the total workforce system. We have a flexible set of arrangements for people to work full or part time or a combination of both throughout their career in the Army. That helps us have a workforce in 157 locations, as some of these are sparsely populated.

Our capacity to leverage our total workforce system means we can leverage our part time brigades to project force from the bases South of the Tropic of Capricorn into the northern geography to reinforce and protect our sustainment capabilities in that part of the country.

We have restructured our 2nd Division to be a division which leverages our total workforce system. Its six formations leverage our part time people in great part.

With recent natural disasters in Australia, such as the bush fires, the Army has been mobilised to help the nation in non-defence crises. This has meant that C2 has been used for national mobilisation as well as to transport equipment to move force to the point of need.

There is the challenge of overtaxing the Army for such tasks, but it does suggest that mobilisation is a whole of nation effort, not simply a tip of the spear warfighting support effort.

We then discussed one aspect of mobilisation which has become clearly evident, namely a relationship between government and industry to provide for war materiel at levels of effectiveness and not just in time efficiencies.

Alan Dupont and others spoke at the 30 March 2023 Williams seminar on the impact of the Ukraine War which has demonstrated the absence of the kind of arsenal of democracy which Australia and the liberal democracies need.

We did discuss the munitions challenge which requires significant investment in development and the buying of weapons stockpiles. With regard to Australia, Lieutenant General Stuart noted:

We need to have the capacity to store, maintain, and perform upgrades on guided weapons in Australia using an Australian workforce. And that is a prerequisite to the capacity to then be able to either provide component manufacture or assembly or actually to manufacture guided weapons and explosive ordnance in the country.

For Lieutenant General Stuart, the Australian Army has a key role to play in the way ahead for the direct defence of Australia and the role of the ADF and the Australian nation in the region. Doing so will take imagination, resources, and commitment—qualities which are always in short supply, at least in my view.

Mobilisation

The way ahead on Australian defence entails a whole of government and a whole of society challenge, notably in terms of how defence had changed with the multi-faceted threats which China poses to Australia and the liberal democratic global order.

To prepare and respond requires a mobilisation strategy. With the wildfire and pandemic crises, Australia faced significant challenges which required mobilisation and broader social awareness.

The demand signal notably on the Army was very high and has impacted on the force. It is clear that the Army being present throughout Australia is well positioned to spearhead such efforts but frankly should not be given the primary task of staffing out the entire effort.

This requires shaping a mobilisation system available for social, natural and environmental crises and available in times of need for national defence. Mobilisation in this sense is not strictly a defence term but rather one which needs to be understood as a broader requirement for how society is evolving in age of various challenges.

And public involvement is crucial to shape the kind of awareness required to have a fully supported approach by the nation to resilience. The need is for robust resilience, not reticent resistance.

In an interview with AIRMSHL John Harvey (Retd) after his presentation at the March 2023 Williams Foundation seminar, we discussed the importance of shaping an effective mobilisation system as part of the way forward for Australian defence. Harvey underscored that what was required in the new context a whole of government, society and whole of alliance capability.

With regard to mobilisation, he made the very sound point that mobilisation was important across the whole of government and society to deal with a variety of challenges, not just defence. Indeed, if one correlated mobilisation simply with defence, that would lead to failure to focus on the much broader challenge which is best characterized by a capability for national resilience.

From this point of view, deterrence then is based on social cohesion and national cohesion to sustain Australia through the pressures which the changing global system puts upon her.

At his presentation to the RUSI Australia meeting in April 2023, John Blaxland provided a briefing entitled "AUKLUS, the DSR and Beyond" in which he highlighted the importance of mobilisation as part of the way ahead. In fact, his next book is entitled *Mobilising the Australian Army*.

The broader need for mobilisation is the engagement and involvement of the citizenry in the protection, health, viability and defence of their island continent. It is not a task for the ADF as a boutique force at the service of the tasks given to it by the Prime Minister of the day.

When one considers the breadth of the shift in defence posed by a multi-domain, multi-specter, and cross domain authoritarian world redefining power such as China, the engagement of the citizenry is not a nice to have capability but a fundamental one.

In addition to the broader set of national challenges, the specific focus on the imperative of creating in the Northern Territory to right kind of infrastructure to support direct defence and "impactful projection" requires mobilisation of national resources for a region with scarce populations and infrastructure to support such efforts.

Glenn Brown and John Coyne made precisely these points in their 28 April 2023 article which they entitled "turning the defence review into action will require a major mobilisation."

Northern Australia's critical transport infrastructure has multiple single points of failure that create strategic vulnerabilities. The logistics and supply chain networks to and from northern Australia are already subject to and stressed by weather-driven disruptions.

Northern Australia's economy has limited market depth and surge capacity, so completing major new works requires engagement across multiple government agencies and market sectors, and awareness of the coming waves of large-scale resource projects in the Territory and their massive impacts on the 'normal' economy and supply chains of people and products.

The DSR mentions tapping into the civilian and mining infrastructures of the north. That's commendable but it means that energy and minerals sector players will become 'interested parties' and must join the planning cohorts.

The NT, like all the states and territories, has a significant housing shortage. Achieving the DSR goals will require significantly large work forces who will need to be housed. This must be addressed as a priority.

Defence will need to assess what the force posture changes in northern Australia will require of the ADF, and state and territory governments, defence industry and the construction community.

In addition to defence and minerals commitments, these demands appear likely to present serious challenges along with the need for facilities for visiting U..S and Japanese personnel. What happens next and how the guidance of the DSR translates into operational plans and tactics will be significant undertakings that all the stakeholders will need to support. This will require the largest mobilisation effort in Australia since World War II.[7]

The general perception that defence is a task for professionals and not of the citizenry has become cemented in all of the liberal democracies with the exception of the Baltic states, the Nordics and the Poles. But this "citizenry gap" is a crucial challenge facing Australia along with its major allies.

As Scott Davidson put it in an article published on 1 March 2023:

Like many Western societies, there has been an increasing gap in Australia between militaries and the societies they represent. Since the abolition of National Service in Australia in 1972 – midway through the Vietnam War – the impact of wars after the end of the Cold war have not affected society directly.

The wars of choice since World War II removed the requirement for mass mobilisation or defence spending. Australian spending on the military through the last 20 years never exceeded 6% of the Federal budget. This minimises the tangible impact on the population of a nation at war; such as increases in taxation, government directed shifts in industry, or national rationing of critical resources.

The enduring nature of war and the approach China might apply to major conflict in the Indo-Pacific means future war for Australia will likely be protracted and costly. Even if the theatres of battle are distant from Australia, the economic disruption from war in the Indo-Pacific would be significant due to the economic interdependence from globalisation. Arguably, the impact on societies would be broader than previous instances of total war in the 20th century, even without considering the threat of nuclear weapons.

The dependence of civilian society on the two new domains of war, cyber and space, means that regardless of how governments may try to insulate their society from the cost of war, there will be impacts on the modern conveniences that people today take for granted. This impact added on top of the other costs of war would exacerbate the vulnerability to Maoist Protracted War, with the Australian population increasingly unwilling to endure a long, costly, drawn-out war which undermines their standard of living.

Australians have lost the experience of privation and national mobilisation that contributed to success in World War II. The current focus on prioritising social comfort over national security and shielding populations from the costs associated with war, undermines the resilience of the population to endure and accept privation associated with an existential conflict. This is natural when the threat is low.

However, the status quo has changed and so should Australian's mindsets. Compare Australia's current posture with the preparations in China of hardened communication systems, reserves of fuel and logistics, and a strong ideology rooted in a national narrative that has prepared the Chinese population for a protracted struggle.[8]

The Nuclear Factor

Overhanging any consideration of the way ahead for Australian defence is the nuclear factor. And it impacts in three different ways.

The first way is the coming of the nuclear submarine to the Royal Australian Navy. Here not only will the RAN receive, then build and operate nuclear submarines, but they will work with the U.S. and the UK and perhaps France in hosting their own nuclear submarines when doing patrols in the Pacific.

As Vice Admiral Mead noted, Australia will build operating facilities for these submarines and have the security systems which the U.S. considers necessary to protect them.

And this means building a base on Australian soil which could raise domestic objections against the critical claims of those Australians who have historically opposed nuclear power and its use on Australian soil, not the least of which has been past historical statements and positions taken by the current Prime Minister.

Divisions within the Labour Party and the role of the Greens in the Australian political system will make opposition to things nuclear even including the AUKUS submarines a factor going forward. And given the nature of the Australian system of federal government will raise questions as well of where the base or bases are located and with what political effect.

Does political division pop up at a later point to threaten the AUKUS submarine deal?

The second way is the crucial question of Australia's ability to deal with energy resilience. Australia has both abundant coal and uranium but policies support diminishing reliance on coal and not use of uranium going forward.

Australia could if it so chose be energy independent largely due to its ability to tap into solar and nuclear energy sources, and with a significant rebuild of the electric power grid. But there seems to be little appetite to go the nuclear energy route for energy generation.

But how will Australia have enough energy sources, most notably fossil fuels, in times of an enduring crisis? And how will it ensure that it has enough merchant shipping to bring supplies into the country during such a period?

To operate their submarines, Australia will build a generation of nuclear power officers. But unlike in the United States or France, these officers will not operate against the background of a broader domestic nuclear power engineering sector. Presumably, domestic capabilities of some sort will have to be created for training. But this is a glass half full when it comes to realizing the benefits of nuclear power generation for an energy independence stance in Australia.

But regardless of a position on domestic nuclear power, Australia and its allies face three nuclear powers in the region, and possessing nuclear weapons does matter in shaping one's posture of operating conventional forces.

That nuclear deterrence affects conventional operations has been obvious in the war in Ukraine. But that war has raised questions about the spillover effect into the Pacific.

I wrote a piece published on 9 April 2023 which highlighted the impact of nuclear deterrence on the sanctuaries of rear support for the war which underscored how nuclear weapons play into conventional operations even if they are not directly used, which is a key consideration for any consideration of ADF use in a conflict with China.

"There are many aspects of the war being studied by militaries worldwide, but one aspect of the conflict is dramatically evident –

the war is fueled by supplies from the West delivered by supply lines from the West into Ukraine and supplies from Russian territory into Ukraine supporting Russian forces.

"There is evident concern to avoid striking either rear area and there can be little explanation other than the fear of escalation why this is so. And the escalation at risk is a threat of nuclear use. Does anyone really believe the Russians would tolerate this if not for British nuclear deterrence reinforced by the American nuclear force?

"As Paul Bracken has noted: "Someone called me yesterday to ask if the Russians would actually use nuclear weapons. My response was "they already have." It's a nuclear head game, and very dangerous.

"The purpose of the Russian nuclear alert is to deter NATO from massing its forces against Belarus and Ukraine borders. And to signal that the U.S. had better not open up a big electronic warfare or cyber campaign to disrupt Russian Air Forces over Ukraine.

"A U.S. or NATO cyber campaign against distributed tactical nuclear and mobile missiles of Russia would manipulate the risk of escalation, which is why Putin ordered the alert."

"But there is a growing danger of escalation. On the Western side, prior to the war, there was a de facto expectation that the West could trade space for time in a Russian attack on NATO. But no state that borders Russia (and now we have a new one in NATO), watching the atrocities the Russians have committed in Ukrainian territory, can want to trade space for time. This could well mean that these border states want long-range strike weapons as part of their arsenal along with more credible air and missile defense capabilities.

"And such long-range weapons can be land or air-based and are being built in the West to deal with the distances in the Pacific to deal with the China challenge. Additionally, the Ukrainians have reputedly asked for longer range missiles which would allow them to hit Russian supply areas deeper into Russia.

"The Russians can contemplate various ways chemical weapons could be used against supply chains operating within Ukraine from the West. Does the West have sufficient defensive capability to keep the supply chain rolling in such circumstances? Does the West have a realistic offensive answer?

"But the core question is simply put: does the possession of nuclear weapons effectively create sanctuaries in your territory in case of conflict?

"Does this work with regard to extended deterrence as well by the United States with its allies?

"Would this apply to the defense of Australia as it expands its basing support for the United States? Does this work as well in the Pacific with China, Russia, North Korea and the mainland of the United States effectively sanctuaries? How does the question affect warfighting strategies, muti-domain or otherwise?

"And most of all, the drive toward expanded numbers of nuclear states envisaged in Bracken's second nuclear age is hardly undercut by the Ukraine war experience to date. It is most likely a positive proof of the need to do so by a major state."

This last point raises the key question of whether other states in the Pacific region will decide to become nuclear weapons states, such as South Korea. Australia could see its defence position complicated by proliferation in the region spawned by events in Europe, Iran, and North Korea and in ways that complicate the credibility of American nuclear deterrence policy.

And that is the elephant in the room – American extended nuclear deterrence. In a piece published in Survival on 4 February 2022, Professor Stephan Frühling co-authored a piece which precisely looked at the necessity and the path whereby U.S. extended deterrence could be enhanced in the face of the Chinese nuclear build-up.

As they noted:

Forward-based nuclear forces are a central element in coupling allied and US security, creating risks of entanglement for the adversary and addressing adversary threats of limited use. NATO has such forces. In the NPR, the US should present for consideration the possibility of forward-basing nuclear weapons in the Indo-Pacific, as well as stationing dual-capable aircraft there, perhaps including some with South Korean and Japanese crews that are certified to carry out nuclear missions.

Given the relative lack of strategic buffer between South Korea and North Korea, and between China and the southern islands of Japan, even relatively short-range dual-capable aircraft systems could fulfil an important coupling role.

As an operational expression of Washington's willingness to give allies some say in the avoidance of nuclear sanctuaries during a conflict, such forces would compel adversaries to take seriously the role of US nuclear capabilities in a major conflict. This dispensation would introduce escalation risks that China currently does not face into any counterforce campaign against the US and its allies.[9]

In the 1980s, I spent an enormous amount of time on this issue, notably in the context of the Euro-missile crisis. I did much research and published many articles and books on the French and British nuclear arsenals and their relationships with what the U.S. might contribute in a theater war.

This subject has returned with a vengeance and simply cannot be ignored when looking at the question of deterrence in the Pacific against a mature nuclear power like China combined with a personal nuclear arsenal possessed by the leader of North Korea or the question of how Russia and its war in Ukraine will affect its own nuclear posture.

This is hardly an historical study, but living history so to speak. How does the return of the nuclear dimension and evolving U.S. policy affect Australian options and ways ahead?

The Challenge of Anchoring Defence for the Liberal Democratic Powers in the Indo-Pacific

Although an island continent which is located at the end of the Pacific, Australia is now deeply enmeshed in the global change affecting the reshaping of the global system.

It has become a central player in how the Chinese way of reshaping global power plays out, whether it wants to be or not. It is significant enough to set its own course but not powerful enough to do so unilaterally.

And it needs a defence policy that fits this period of global upheaval. In my view, it is by becoming more resilient and doing so by being interactive with and driving change amongst its allies to shape credible paths to more resilience in the face of the authoritarian powers that it provides a leadership role.

I discussed this shift with Ross Babbage during my April 2023 visit. Here, he highlighted what he considers to be three key elements in Australia to being able to shape a broader defence capability. For Babbage, shaping a broader defence capability is not just about the ADF and its own operational capability. "If you're looking at it from Beijing's point of view, they'd have to think very carefully about messing with us for we do have a very capable although small military and we have even more powerful friends." But the ADF lacks strategic depth and sustainability. As Babbage noted: "We are in danger of being a one-month operational military in case of conflict due to the lack of economic and industrial depth, such as the provision of fuels and key munitions and spare parts."

The second aspect for Australia is its alliance structure. As Babbage underscored, Australia has focused upon ramping up its alliance working relationships to the point where its own forces can more effectively integrate with the Americans and are working towards greater cooperation with other allies as well, notably the Japanese.

The result is clear: "The sum of alliance efforts is greater than any of the parts. This is a consideration which Beijing has to realize is not working to its advantage. The Chinese threat has drawn many nations in the Pacific closer together to resist authoritarian interference."

And it is not just about Pacific allies: a number of European states have woken up to the realisation that China directly threatens their interests, and they have to find ways to contribute to the deterrence of China as well. Babbage noted: "We will cooperate with a range of others, including a number of relatively powerful and capable Europeans with whom we have long-standing partnerships."

And that led to the discussion of the third element in Australian deterrence, developing more effective regional partnerships. Here he discussed evolving relationships with India, Indonesia, and other Southeast Asian and South Pacific countries. Australia is working hard to develop closer military, security, economic, technological, and diplomatic relationships that can strengthen regional cooperation and deterrence. Working with its neighbourhood much more directly and effectively is a key part of shaping the way ahead for Australia's deterrence strategy.

The challenge is to move from the near and midterm efforts at enhanced national military capability and allied interoperability to a stronger capability for resilient societies that empower enduring forces, not just one-month militaries. The close allies need to review and restructure their strategic supply chains as a matter of urgency to reinforce each other's economic and industrial strengths and cover their respective weaknesses.

New levels of allied cooperation are required along with new planning and management mechanisms. These initiatives are needed urgently if the allies are to have a credible deterrent going forward and if they are going to be able to endure and sustain themselves in the event of a major conflict.

Australia's ability to enhance its ability for the direct defence of the continent while working with allies to operate as a sanctuary in times of crisis is the key way ahead.

THREE

The Case of Australian Maritime Strategy

A key case study of the challenge of shaping a new approach to Australian defense revolves around shaping a realistic an effective maritime strategy. This was the focus of the first of two Sir Richard Williams Foundation seminars in 2024. The seminar was held on 11 April 2024 at the National Gallery of Australia in Canberra.

The focus of the seminar was identified as follows:

The strategic environment continues to deteriorate on a global scale with Australia's immediate region the source of increased risk on a number of levels. Meeting preparedness requirements and implementing the Defence Strategic Review (DSR) while building for the future will place significant strain on the Defence enterprise.

In meeting these near term and longer-term needs, one aspect of the environment endures: Australia's strategic geography, which demands a resourced, coherent, and executable maritime strategy.

In short, a sophisticated and credible maritime strategy is a multi-domain, multi-agency, whole of nation effort requiring an enduring focus on the generation of national power and options that contribute to Australia's national security outcomes at the lowest political risk.

But a maritime strategy considers more than naval operations. It details the ends, ways, and means necessary for the generation of national power that serves all of a nation's interests.

A maritime strategy must therefore contribute to other elements of national power such as diplomatic, informational, and economic, and is not enough to focus on military capabilities alone.

It must address the broader efforts of Defence and the other agencies which contribute to the security of borders and the exclusive economic zone, as well as protecting the mobility of trade and data either on, above or below the surface.

In many ways, therefore, the objectives of an Australian maritime strategy are no different to other nations, especially those that also rely heavily on the oceans for the passage of trade and the development of economic power.

However, the vast area of interest and Australia's relatively small population poses a complex challenge when identifying the ways and means by which those national objectives are achieved.

In a practical sense, a maritime strategy requires a highly integrated, multi-agency, multi-domain response enabled by, among others, connectivity, logistics, bases, stores, and decision-making superiority. And with an increasingly challenging threat environment, this must all be resilient and ready.

Overview of the 11 April 2024 Seminar

Last year's DSR highlighted the ramped-up threat to Australia and the need to focus on the region, its partnerships and a more effective defence effort by Australia in the regional deterrence context.

The focus of the government in its subsequent priorities has tended to focus on longer term acquisitions, first in terms of nuclear submarines through the AUKUS relationship and for a new surface fleet in its recently released surface fleet review.

A multi-domain operations discussion builds on the work of the Foundation since I have been writing the reports since 2014 upon building a fifth-generation force, which after all revolves around sensor-shooter relationships built across an integrated force delivering multi-domain effects or what I prefer to call a kill-web enabled force.

The focus is upon how you get full value out of your force now and to build out that extant force in the future to become more lethal and survivable. If you are focused on the fight tonight, which any credible combat force must focus on, then long range assets are projections of the possible, not augmentations of the credibility of the operational force.

So any multi-domain discussion inevitably focuses on the way ahead for the force in being, rather than a force planning discussion of a projected future.

When you add the key consideration of what that force is in support of achieving, inevitably gaps are identified, and the question then is how do you close the most significant gaps which threaten your security and defence interests.

This perspective then raises the question of the means to the end of what one might consider a maritime threat envelope and strategy to deal with that envelope.

In other words, one would expect the seminar discussion to focus more on the transition challenges of the ADF and the nation to deal with threat environment in the near to midterm rather than in 2040.

That is what happened at the seminar in which speakers started by highlighting the importance of focusing on the here and now rather than on the force that might exist in 2035 or 2040.

After the initial presentations focused on the current challenges and the role of the ADF and the nation to prepare to deal with them, the discussion shifted to whether Australia had a maritime strategy and if so what were the priorities of such a strategy.

The majority of the presentations focused on specific service or industrial perspectives with regard to how best to meet the multi-domain requirements for the evolving Australian defence challenge.

But at the heart of the discussion was really the major challenge facing Australia: how to close defence gaps? How to engage the nation beyond the ADF in the broader defence challenges facing Australia? How to build a sustainable force?

How does the ADF get more capable in the next three-to-five years and to do so in a way that is a prologue to the anticipated force transformation being designed?

Peter Jennings was the first speaker and he underscored that the

DSR had highlighted the near-term threats but the investment was in forces a decade away.

He put the challenge as follows:

Governments can and do promise to spend unbelievable quantities of money on the future force but you only know what you get when you open the box.

Not one cent of it buys deterrence today.

From a deterrence perspective there is potentially some risk in promising strong deterrent capabilities in the future while maintaining the military capabilities of a skinned cat in the present day.

That is the risk of pre-emption. Indeed, one reason why analysts are so worried about a mid- to late-2020s risk of conflict against Taiwan, or in the South China Sea, is that Xi Jinping may calculate that he faces a 'use it or lose it' choice with the PLA.

Xi's best chance of strategic success to achieve unchallenged military dominance in the Pacific are maximised by early action before his opponents' next generation military capabilities are realised and while the democracies are internally distracted and divided.

The tragedy is that there is so much which could be done with a bit of political and Defence push to strengthen ADF and national capabilities in the relative short term.

For example:

- *Ramping up domestic ammunition production and stockpiling.*
- *Establishing offensive drone capabilities on the basis of existing technology – not everything has to be quantum, AI, hypersonically joint and enabled.*
- *Funding some of the incredibly smart military capabilities that have been developed by Australian businesses.*
- *Researching some of the remarkable military and operational achievements which the Ukrainians (with allied help) and the Israelis have used in recent months.*

Here I'm not just talking about drones; but also optimising air defence capabilities; integrating intelligence and battlefield situational awareness; finding the right balance between exotic and more prosaic technology; working out how to get things in production in less than a decade.

There is so much that could be done, so much so, in fact that our failure to do any of this makes me wonder if it is not the case that the government and Defence establishment is actually getting what it really wants?

The second presentation was by Mike Pezzullo, the former Secretary of the Department of Home Affairs, who made an impassioned speech reminding the audience that building an effective defence structure is not simply the task of the ADF.

The society needed to be engaged in shaping an Australia more capable of defending itself. You cannot outsource defence and security to an alliance or to the professional military for one needs to build a more resilient and sustainable Australian society and nation.

Challenges Facing Australia in Shaping an Effective Maritime Strategy

Let me now turn to the presentations by Parker and by Commissioner of the Australian Border Force, Michael Outram for they provided a significant baseline for the discussion and for any serious analysis of the way ahead in the maritime dimension for Australia.

A discussion of how multi-domain operations could enable Australia to more effectively execute an effective Australian maritime strategy pre-supposes that Australia has a maritime strategy and a fairly clear sense of what its maritime interests are which need to be protected.

The government's Defence Strategic Review last year and the recently released defence strategy certainly highlight a range of maritime capabilities which the government has focused upon to determine how best to enhance Australian defence.

But what are the tools in place and the new tools which need to be acquired to enhance Australian maritime security and defence?

Or in other words, there are prior questions to the question of acquiring new ships.

- What are the threats?
- How to organize for them?
- Who should be responsible for dealing with them?

- And how to deal with them most effectively with what means?

The government has highlighted a central focus on a strategy of deterrence by denial. But if the Chinese seriously disrupt Australia's ability to move goods by sea, who is denying whom?

At the seminar, two presentations directly dealt with the questions of maritime strategy and security.

The presentation by Jennifer Parker of the National Security College of the Australian National University addressed the question of whether Australia actually has a maritime strategy and if they did what was it?

The second was by the Commissioner of the Australian Border Force, Michael Outram, and dealt with the very significant question of the daunting challenges to maritime security in a period of disruption of the "rules-based" order.

Jennifer Parker speaking at the Williams Foundation seminar on April 11, 2024.

Parker provided a broad stroke analysis of maritime strategy, rather than reducing the discussion to what platforms and capability which Australia has to operate in the maritime domain. In her presentation she defined maritime strategy for Australia as "the plan to protect Australia's maritime strategic interests using all aspects of national power."

Her perspective meant that she would conclude that a maritime strategy defined as deterrence by denial would be too narrow to capture the full spectrum of demands from the maritime domain that required an appropriate security and defense regime to determine and defend Australian maritime interests.

She mentioned several cases of conflicts in the maritime domain which have been evident in the recent past which illustrate the broad nature of the challenges to be dealt with.

One was the targeting of shipping to send a political message which is evident in what is going on in the Middle East. Given Australia's dependence on maritime trade, this is a problem which Australia clearly needs to be prepared for.

The second has been evident in both the confrontation in the Black Sea and the challenges being addressed by the Nordics and the Baltic states involving the Russians and the Baltic Sea. This is a question of port security and undersea cable protection. Here one is talking about active measures for security and defense, not simply posturing for deterrence.

The third has been the importance of "information war" in the maritime domain evident in the Chinese anything but gray zone confrontation with the Philippines. The Philippines are pulling the strings on their alliance relationships to generate defence options, but they have used transparency to fight back in the information war with the Chinese.

It is also the case that they are adding new defence capabilities which will allow them to counter directly Chinese aggression which again is not building a posture for deterrence by denial – it is about directly confronting the adversary, which has been a major failure, in my view, of characterizing the Chinese as operating the gray zoos.

In a book review I wrote about a book dealing with China and the gray zone, I underscored the limitations of using this concept from the standpoint of shaping credible action policy:

This is how I highlighted the challenge:

"Western analysts have coined phrases like hybrid war and gray zones as a way to describe peer conflict below the level of general armed conflict. But such language creates a cottage industry of think tank analysts, rather than accurately portraying the international security environment.

"Peer conflict notably between the liberal democracies and the 21st century authoritarian powers is conflict over global dominance and management. It is not about managing the global commons; it is about whose rules dominate and apply. Rather than being hybrid or gray, these conflicts, like most grand strategy since Napoleon, are much more about "non war" than they are about war. They shape the rules of the game to give one side usable advantage. They exploit the risk of moving to a higher intensity of confrontation.

"What limits should be crossed to manipulate the risk of going to a higher intensity of competition?"[1]

In our period of history, no credible defence approach can be designed without a strong security foundation. It is not simply about the point end of the spear and when you use it, or the forces generated in a force design for the professional military. It is about having a society and economy built on solid foundation of security.

The presentation of the Commissioner of the Australian Border Force provided a broad understanding of the need for a robust security policy to underwrite a credible maritime strategy. Michael Outram highlighted the importance of Australia's maritime domain citing its $1 trillion in annual trade and 5% of GDP.

He underscored that there are wide ranging security threats in maritime domain which include illegal fishing, cyberattacks, and biosecurity risks. The challenges of maritime security in the Indo-Pacific region, include the resilience and agility of criminal networks and the limitations of publicly accountable bureaucracies.

To deal with these challenges. the Australian Border Force (ABF) is collaborating with Pacific island nations to build capacity and address growing criminal threats, including illegal fishing and migration.

Commissioner Outram speaking at the Williams Foundation seminar April 11, 2024.

The Australian Border Force (ABF) and Defence have an overlap in their missions, particularly in the maritime domain, but this overlap is not static and can shift depending on circumstances. The Maritime Border Command within ABF working with defence focus on the challenge or surveilling and monitoring vessels operating in Australia's maritime domain.

In my own view, there is a significant opportunity to leverage autonomous systems into an integrated security and defence operational culture which will be critical in order to be able to deal with the larger issues which Parker highlighted.

The Commissioner went on to argue that the time was ripe for from serious rethinking about how the Australia government needs to work in this area.

He identified several areas where progress needs to be made:

- Consider developing a new civil maritime security strategy that addresses strategic coherence, governance, funding structures, and the definition and scope of civil maritime security in light of changing geopolitical and technological conditions.
- Conduct a series of future focused scenario-based planning exercises to evaluate whether the current operating model or an alternative model such as an independent Coast Guard could be more effective in addressing strategic shifts over the next decade.
- Give serious thought to whether the regional security situation, shifts in technology, and other factors require a different strategic approach to civil maritime security and a redefining of the scope of operations.
- Determine if the current civil maritime strategic architecture, planning, governance, funding, and structure remains fit for purpose over the next decade.
- Assess if the civil maritime operating model of the past 20 years can be sustained and remains fit for the strategic purpose over the next 20 years.

He concluded with the importance of addressing long-term structural funding issues to maintain a fit for purpose civil maritime capability appropriate to Australia's interests.

The two presentations taken together underscore the need to focus on how the Australian government is organized to address maritime security and defence issues. And I would argue that to use new technologies in this domain is also required fundamental organizational change as well.

FOUR

General Considerations for an Effective Strategy

What does a 21st century defence strategy look like for Australia in a multi-polar authoritarian world?

The answer is that it does not look like the defence strategy which has been followed throughout most of the post-war period.

The threat envelope is quite different. There is no American and Western managed rules-based order dominating the world. There are diverse authoritarian movements and states which follow their distinct interests but play off of one another.

As one analyst has put it: "But the end of the Cold War has led to the atomisation of threats – many of these threat groups possess weapons and backing from powerful regional states that in some cases make them as capable as state-based actors.

"Nowhere is this more apparent than in the Middle East, where improved military capabilities are combined with an ideological zealotry that makes normal cost-benefit calculations underpinning deterrence redundant. This makes it very difficult for Washington to achieve the type of deterrence on which long-term regional stability is often based."[1]

And the direct threat to Australia is broad and not narrowly focused on what the Australian Defence Force can do. A sustainable force and a resilient Australia are beyond the scope of narrowly considered defence investments in a ready force. They are all of government and all of society challenges.

The Need for a New "War Book"?

At the Williams Foundation Seminar held on April 11, 2024, the former Australian Secretary of Home Affairs, Mike Pezzullo, clearly underscored how different the era into which Australia and its allies had entered compared to the previous one.

Mike Pezzullo presenting at the Williams Foundation April 11, 2024 Seminar.

As he put it in his presentation:

"What might this mean for Australia and specifically the Australian defence enterprise?

"Defence planning is rightly focused on a wide range of contingencies. With very little notice the Australian Defence Force could be called upon to undertake rapid deployments into the nearby arc of small states.

"While necessary and important, such ventures would only be marginally relevant to today's great issues of war and peace. The same could be said of vital operations in support of distressed communities in the wake of natural disasters.

"Given long lead times, defence also has to focus on complex capability and programming issues, especially as related to the planned force of 2035 and beyond."

But he cautioned that the threats in front of Australia now needed to drive a re-set in efforts that considered the engagement of the society in its own defence, not just crafting hypothetical future force structures.

And he quite correctly warned against the danger of shaping Potemkin long range capabilities that may never arrive in time to make a difference.

He focused much of his attention on the need to engage whole of government in working with economic leaders in shaping a way ahead for a more resilient Australia that could support a sustainable ADF along with core allies working with Australia as a strategic reserve both to deter and to prevail in crisis situations.

He underscored: "The most important question is whether a nation at large has the structures, capabilities and above all, the mindset and the will, that are required to fight and keep fighting to absorb, recover, endure and prevail. These cannot be put in place or engendered on the eve of the storm.

"Now as a practical suggestion to focus relevant effort, we should consider modernizing the earlier practice from the 1930s and and then again from the 1950s of the preparation of a war book. The war book of those times were guides on what would need to be done and by whom, in the event of war. Preparing a new war book would help to focus the national mind."

He clarified his suggested approach as follows:

"A new war book would deal with the entire span of civil defense and mobilization which would be required to move to a war footing, consisting of a range of coordinated plans. Some would deal with critical infrastructure protection, and national cyber defense. Other plans would deal with the mobilization of labour and industrial production covering supply chains, industrial materials, chemicals, minerals, and so on.

"Sectoral plans would address the allocation, rationing and or stockpiling of fuel, energy, water, food, transport, shipping, aviation, communications, health services, pharmaceuticals, building construction resources, and so on and so forth.

"They would also be plans for the protection of the civil population covering evacuation, rapid fortification and or shelter construction, and for augmenting police fire, rescue and ambulance capacities, and also dealing with social cohesion, border security, domestic security and public safety.

"Lessons could be adapted from international experience, especially Ukraine and Israel, as well as from domestic experiences such as natural disasters, and the COVID pandemic noting however, that war is different."

In short, 21st century defence is not narrowly focused on the ADF and long range investments in a future force.

All one has to look around you and find the activity of the multi-polar authoritarian world and the end of the American-led "rules-based order" to understand the future is now.

How best to shape a way ahead in terms of augmented capabilities in short to mid-term and engage the nation in its own defence for the longer term is really the challenge.

Managing Trade-Offs in Force Structure Development

When a nation is facing a deteriorating threat environment, one key challenge in ramping up defence investments is how to balance enhancing the current fight to night force with new future platforms as part of a future force structure.

This problem is compounded by the changing nature of the threat envelope for the liberal democracies.

They now face a multi-polar authoritarian state and movement threat envelope whereby these states play off of one another and have various kinds of working relationships which fall short of a complete alliance, but together generate a diverse and diffuse threat to the liberal democracies.

And when it comes to information war, they have a huge advantage of access to the social media-dominated world provide by liberal democratic systems compared to the face recognition controlled authoritarian regimes.

Peter Jennings presenting at the Sir Richard Williams Seminar April 11, 2024

But there is another challenge as well facing force structure design.

The most dynamic new systems for innovation are software designed and AI enabled systems which simply do not follow the pattern of developing and procuring legacy platforms. If you don't use maritime autonomous systems, for example, you cannot redesign them for you do so in direct relationship to their use.

And as your current force becomes a hybrid one with the growing input from autonomous systems, what then is the nature of the future force which one designs based on legacy thinking?

The challenge of the tension between dealing with growing threats now and delaying design responses much later was highlighted in Peter Jennings, Director of Strategic Analysis Australia, presentation to the recent Sir Richard Williams Foundation Seminar held on April 11, 2024.

The main thrust of the presentation was Jennings perceiving a significant gap between the government's emphasis on the near-term threat and its defence investments. The Australian government is not dealing with ways to enhance ADF capability in the near term but putting their priority investments into a future force.

Jennings noted:

Our worsening strategic outlook is a constant theme in Defence Minister Richard Marle's speeches.

Here is Mr Marle's comments at the Sydney Institute on April 4:

"Recorded military spending in the Indo-Pacific region has increased by almost 50 per cent in the past ten years, with China engaging in the biggest conventional military build-up in the world since the Second World War.

"In the year 2000, China had six nuclear-powered submarines. By the end of this decade, they will have 21. In the year 2000, China had 57 major warships. By the end of this decade, they will have 200.

"These investments are shifting the balance of military power in new and uncertain ways. We are in an environment where the risk of miscalculation increases, and the consequences are more severe.

"And as China's strategic and economic weight grows, it is seeking to shape the world around it.

"For a country like Australia this represents a challenge."

In these comments Mr Marles is absolutely right.

If you don't understand that Australia is facing an increasingly threatening strategic environment, one where the risks of war in the mid-2020s is substantially growing, well, either you must be paying no attention to international developments, or you might conceivably be working in DFAT (Defence Foreign Affairs and Trade).

But what has been the practical response according to Jennings?

"The more our governments seem to talk about strategic risk, the less it seems that we are actually able to take practical steps to strengthen the ADF to present a deterrence to conflict."

In his presentation, he ends by highlighting the impact of investment in the autonomous systems technologies which Australia already has access to and has experimented with. Indeed, one of the great ironies is that Australian industry has contributed significantly to Ukrainian defence efforts in various forms of air and sea autonomous systems, but has not applied this technology to the operational ADF.

Here is what Jennings emphasized:

Australia really should engage in a crash program to field an array of drone technology relevant to the maritime domain. There is existing capability available — including Australian proprietary IP which we could bring into service this year or next.

Imagine how motivating for Defence and industry it would be if the Government said there was a billion dollars available for the rapid development of TRL level 9 -- System Proven and Ready for Full Commercial Deployment —drones.

The challenge would be to have fielded capabilities in 2025, let's say before the next federal election.

Impossible I hear you cry?

The Ukrainians are doing it every week.

Our enemies — everyone from the PLA through to the other authoritarian powers, organised crime and the people smuggling cartels — these groups show themselves to me more agile and faster technology adopters than we are in Australia.

We need to think fast and laterally about how to respond. By definition that means current policy processes in Defence are not well adapted to this task. Not fit for purpose as the DSR said.

Hopefully this conference will be able to surface some new and creative ideas for Australian maritime strategy and that those ideas will get a fair hearing.

I would note that a clear example of what Jennings is talking about is what is happening in the context of Nordic integration.

And when one looks at recent Norwegian decisions to ramp up its defense budget and to spend it on programs already being built, one gets the idea of what is possible for a focus on enhancing the current force rather than pushing investment into a conceived of future force.

Notably, several years ago the Norwegian Ministry of Defence worked with the German government on building common procurement of a German submarine. The Norwegians are putting forward more money to build out this program, rather than putting that money aside in a future design build.

Jennings highlighted a crucial question: How do you ramp up ADF capabilities now? And I would add, how do you do so in a way that is a building block for your future force?

It is not about putting money in a drain hole: it is about pump priming the process of improving your fight tonight capabilities and building towards a more capable future force.

FIVE

Air Power in Australia's Maritime Strategy

Airpower is a crucial element of any credible maritime strategy for Australia given the size of the territory, the maritime operating areas, and the need for speed and range to deal with pop-up threats or measured projection of power by the PRC or Russia in the Indo-Pacific or in concert.

Two presentations at the seminar focused on the air power dimension. The first was by Chris McInnes and the second was by Air Commodore Ross Bender, Director General for Air Combat Capability.

The Presentation by Chris McInnes

McInnes highlighted air power's ability to provide rapid engagement and could do so over extensive operational space to deliver desired effects. He argued that in times of an effects-based approach, airpower transforms the time and space dimension for Australia's maritime strategy.

Airpower provides cost-effective options for Australia's national security and cost-effectiveness should be prioritized in Australia's maritime strategy of denial, focusing on delivering large amounts of high explosives to hard targets like warships, airfields, and ports.

Indeed, his presentation was an argument that airpower provided a cost-effective way to deliver massive firepower at range.

His analysis led to his argument that airpower gives Australia time and space to plan, act, and move effectively. This means that prioritizing investment in air superiority to avoid second-best hand in high-stakes situations is crucial.

The presentation can be broken down into three core efforts.

Chris McInnes presenting at the Williams Foundation Seminar April 11, 2024.

The first was to look back at World War II and examine airpower's key role in the Pacific campaign. It played a crucial and decisive impact on the enemy prior to any other means to encroach on the Japanese advances in the Pacific. A combined arms campaign was necessary to recover territory seized by the Japanese empire, but air power was the tip of the spear and a core element of the ability of the allied air forces from sea and land to destroy enemy forces.

The second revolved around the question of the time-space functionality of airpower. Every platform in the joint force is a time-space entity with core characteristics which define what it is able to do. Airpower can move at speed and range no ship can; ships provide slower moving capabilities which can build out a presence force.

As he argued:

"We can swiftly respond with airpower across huge distances with different options in different places on different days. We have more options available and more time in which to consider them.

"But it works both ways. Three hours from Darwin is also three hours to Darwin. PLA airpower can and does hold Australia and its assets at risk across our region in a discretionary, scalable and sustainable manner and in hours, not days. It has already disrupted Australia's sense of time and space. We are inside our warning time.

"I don't think we've quite latched on to what that means though. Airpower shapes how we sense and exploit time and space, which is the most precious thing for Australia and its maritime strategy."

He used a chart to visually underscore the time-space point about airpower.

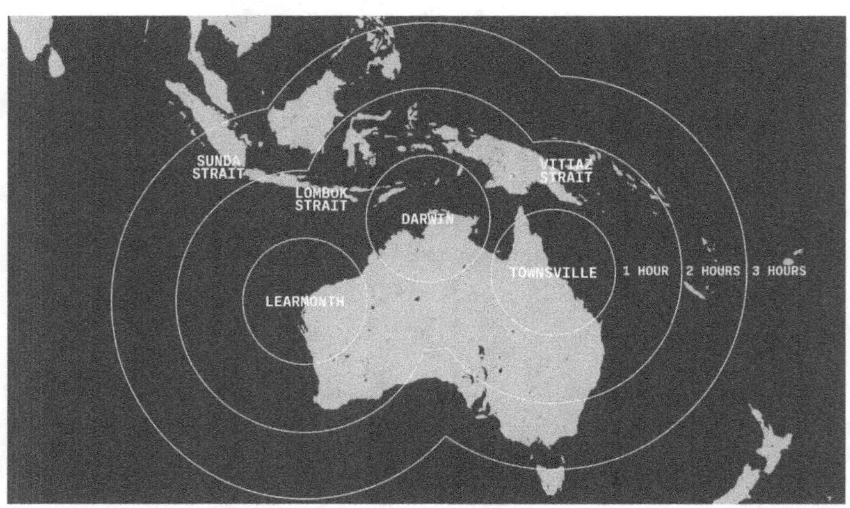

McInnes carefully examined the cost-benefit of weapons delivery enabled by airpower with standoff weapons from sea or land.

He introduced his analysis as follows:

"My analysis is limited to strike as the central operational feature of Australia's maritime strategy of denial. I see the delivery of large amounts of high explosives as determining strike effectiveness and war, and credibility in circumstances short of war.

"Australia's maritime strategy of denial depends on our ability to deliver large and concentrated amounts of high explosive at long range, we could call this impactful projection. We need to hit hard enough to stop movement in different places on different days across a huge area over and over again.

The charts he showed highlighted the range, unit costs per weapon, and warhead class correlated with the launch platform to assess cost effectiveness of ADF weapons.

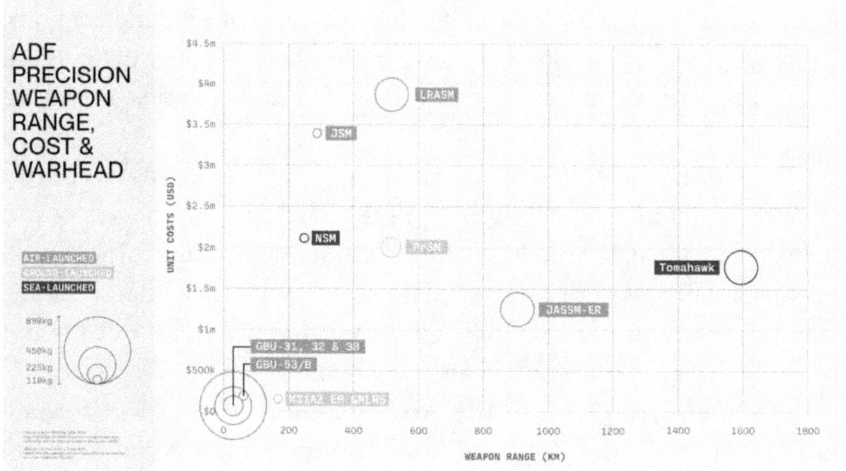

He described the charts this way:

"Unit costs are shown in U.S. dollars and are based on U.S. budget figures going back to the 70s. The unit cost of new weapons will fall as more are purchased.

"The charts clearly show that the delivering the weight of explosive our maritime strategy needs is going to be very expensive, particularly if we become overly reliant on stand-off missiles rather than stand-in weapons in the bottom left corner. It is remarkable how often one reads of the ADF need for long range missiles because of the apparently short range of our air power.

"We must however distinguish between stand-off range – which is the distance a weapon travels from its launcher, and which is what the first chart shows – and effective reach, which incorporates the distance the platform and weapons can rapidly cover.

"When considering effective reach rather than stand-off range, the picture changes dramatically. Stand-in weapons suddenly become some of our longest-range options.

"The second chart incorporates a modest strike radius for the Super Hornet, our shortest-range weapon carrying aircraft. The ADF certainly does need stand-off weapons as they have specific utility against particular targets including air defenses, but they are expensive and inefficient high explosive delivery devices.

"Every exquisite component is single use and many many missiles are needed for strikes, particularly against defended targets. They must carry and do everything internally, including propulsion, navigation and communication. This forces trade-offs, often in warhead size."

"Stand-in weapons are much lower cost and almost entirely warhead, including our largest options. They do rely on expensive delivery platforms, but these are reusable, and multi-role. We do need standoff weapons for specific tasks. But once that is done, stand-in weapons are our most economical and among our longest range options for maritime strategy of denial."

He then focused on the key question of the operational infrastructure for the ADF and its operations, arguing that criticism of airpower as too dependent on vulnerable bases and supply lines overlooked the reality that these dependencies could not be avoided.

This is how he put it when looking at the opportunity costs of different operations:

"What are the trade-offs?

"It seems unavoidable that Australia will always need bases and supplies in its north for military operations in our region. Because at some point, all operations need bases and they will all need air power of some kind. Suggestions that dispersing Australia's assets throughout the archipelago to our north can somehow minimize these costs are hard to square.

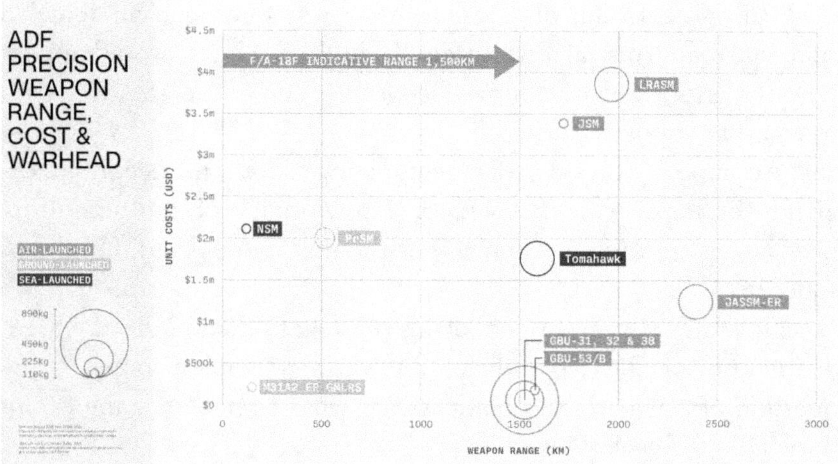

"Even assuming we hold permission to fly missiles through our neighbours airspace, the units will need to supply and defend themselves locally against air and other attack and they will still need supply lines back to Australia, which will have to be secured using air and sea power."

McInnes's closing point was to call for a renewed emphasis on the primacy of air superiority in airpower thinking and investment. As he said:

"However, we will have no options at all without air superiority. And this I contend is where we have reason for concern. In its simplest sense, air superiority is the condition under which we can operate free from prohibitive interference by the enemy.

"Air superiority can be general or local in time and space, it is almost never absolute, and it is a continuous struggle. It is deeply ingrained in the design and operation of Western societies and military forces, including the ADF. It is fundamentally why Australia has an Air Force. It was explicitly the prime campaign for Australian air power until the turn of the century.

"But the Western bloc has lost sight of this primacy over the last 30 years due to complacency and distraction. While the U.S. is reinvigorating it's air superiority approach, its Air Force is struggling for funding while operating its oldest and smallest aircraft fleet since it was formed.

"Russia's invasion of Ukraine has given European air forces a rude wake up. Australia has strengths in the air but it would appear requirements exceed resources geographically and across missions. Mass and tempo are limited."

"Air superiority is a fast-moving competition and deeply unforgiving for those who fall behind. The primacy of air superiority needs to be restored, particularly as the threat grows and funding is squeezed."

The really decisive aspect of his presentation and indeed what is at the core of the evolution of 21st century combat forces, is the question of payloads and platforms or what I refer to as the evolution of the kill-web force.

At the heart of the evolution of fifth generation enabled operations is a significant shift in terms of the sensor-shooter relationship whereby the weapons to be fired at an adversary do not necessarily come for the platform which has the sensor which has identified the target.

This is at the heart of the F-35 development which frankly is still not fully understood and comprehended in the defence analytical world.

If your goal is to deliver lethal payloads, there are a variety of ways to do so.

But at the heart of the issue is where are they launched from and determining what target sets determine which weapons you need and their range. With manned and uncrewed air assets, one significantly reduces the range of the weapon necessary to strike a target as opposed to being launched from land or a ship.

The U.S. aircraft carriers have combined speed, mobility, and launching airpower to close the distance for the missiles being fired.

To conclude, I want to build on McInnes's focus on the need dramatically to reduce the cost of the weapons being used. I would argue that we need to build the functional equivalent of the 155mm shell used by the artillery for an air-launched missile which can be produced across the allied forces.

This will not be a super long missile, probably in the range of 400 miles, but the long range TLAMS which go further are expensive and in limited supply. What this means is that the future belongs to the common air missile produced in quantity that could also be fired from the ground or sea. The functional equivalent of the role of the shells of the 88 in the German army in World War II is what I envisage.

The Presentation by Air Commodore Ross Bender

At the April 11, 2024 Williams Foundation seminar, the former head of the Air Warfare Center and now Director General for Air Combat Capability, Air Commodore Ross Bender, addressed the way ahead for the RAAF in dovetailing with the new strategic focus of the Australian government.

Air Commodore Bender noted that the RAAF although closely partnered with other allies is focused on "conducting campaigns directed to the operational and strategic goals supporting national defense."

It is focused in this sense, and increasingly on the region.

The speed and range of airpower is an essential contribution to the defense of Australia's maritime interests.

As Bender put it: "The ADF must be able to operate across great distances to assure the security of our economic interests and be able to support our allies and partners. Air capability is vital to the maritime domain by providing the speed and responsiveness which it can deliver."

He provided a slide which reminded the audience of an aspect of the range and focus challenge.

He commented on this slide as follows: "And though we'll discuss northern approaches, we should not forget the south with the Antarctic Treaty in mind, which from 2048, any of the parties can call for review.

"I also flag our contributions to some long standing and some relatively new maritime surveillance operations throughout our region, supporting the Australian Government and importantly, our regional partners.

"You might be aware of the Australian P-8 that recently visited

La Réunion. Australia is a maritime nation and the ADF must be able to operate across great distances to assure the security of economic interests and be able to support our allies and partners."[1]

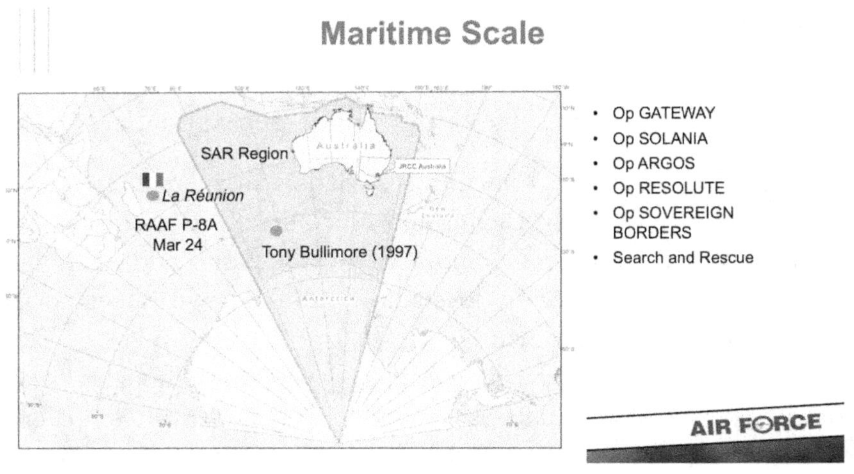

I would note that the ADF is truly dependent on what the RAAF can do as it provides both the air capability associated with the USAF in the United States as well as what the U.S. Navy provides for the U.S. military. It delivers strike, reconnaissance, maritime ISR and targeting data to the ADF.

If the RAAF is not capable of performing its air delivered 360 degree capabilities, then the entire maritime domain defense enterprise for Australia is severely weakened.

In his talk, he discussed the need for the RAAF to develop its own version of agile employment which largely will evolve over ways to operate from the Northern areas of Australia where there are significant infrastructure and work force limitations. The challenge of fuel and logistical support to a distributed force is a major one to be met.

I would note that it has been announced that there is to be acquisition of AGM-158C LRASM anti-ship missiles to be carried on F/A-18Fs, P-8As and eventually F-35As, as well as AGM-158B JASSM-ER air-to-ground missiles. Another item is integration of the Kongsberg Joint Strike Missile on the F-35A. E/A-18G

Growlers will receive 63 AGM-88E AARGM-ER missiles for attacking radars.

And as McInnes noted in his presentation, the range of these missiles in terms of effective attack is expanded by the operation of the air platform themselves.

Air Commodore Bender presenting at the Williams Foundation Seminar April 11, 2024.

Bender then discussed the coming of Triton to the ADF.

"Triton will operate from RAAF base Tyndall in the Northern Territory and be controlled from RAAF base Edinburgh in South Australia, a clear example of the new paradigm for the ADF and the Air Force. The platform is high cost, requires a highly skilled workforce to operate and maintain, but its capability is ideally suited for constant observation of our northern approaches."

But the plan is to expand over time autonomous capabilities augmenting the manned and remotely piloted combat force.

Air Commodore Bender underscored: "Advanced autonomous concepts and capabilities, such as collaborative combat aircraft, can expand the projected envelope of high value, air or maritime assets, while extending their effective reach."

A challenging and I personally believe costly effort that is not fully recognized in realistic budget discussions is simply adapting the RAAF to new operational conditions and contexts.

This is how Air Commodore Bender put It: "There must be important efforts to address a challenge in operating force in Australia. We can't consider our bases as sanctuaries anymore, disconnected from the support base in Australia. How do we continue to operate and demonstrate resilience and maintain the initiative to support deterrence?

"The Air Force is adopting an agile operations concept of a maneuver across a dispersed and hardened network of bases. Of course, this approach must include the measures we can take through the development of integrated air and missile defense capabilities. This protection also requires an understanding of own force signatures, and the automated threat environment, including to supporting and enabling elements.

"An agile posture increases deterrence by being strategically predictable, but operationally unpredictable. Strategic predictability comes from ensuring potential adversaries are left under no doubt about our resolve to ensure survivable, resilient, and enduring airpower operations. Agility at the level we think necessary requires new approaches to combat support, logistics and command and control.

"At its heart, an agile operations concept provides a network of air domain access points to enable aircraft to move rapidly to enable us to aggregate effects, and then disaggregate and reconstitute to complicate advisory targeting. Agile operations enable the resilience of our airpower."

But what is the challenge in moving ahead with such a vision?

What follow are my own thoughts and not those of any speaker during the day of the Williams Foundation seminar from the ADF.

The reality is that the government is cutting airpower in favor of its investments in the future maritime force, notably SSNs and the future surface fleet. This leaves clear gaps with regard to the enhancing of ADF capability in the crucial three-to-five year period facing the ADF.

Government documents and officials have embraced the notion that Australia's warning time is significantly reduced but the reality is that the government is cutting current capability to pay for a force 10 years away.

One needs to be clear.

The decision to cut funding for the fourth squadron of F-35s is a significant reduction in capability. Notably, when one considers the range at which F-35s operating as an allied fleet can move data for targeting, eliminating the numbers of aircraft have an impact.

And the RAAF F-35s are capable of integration with those of the USAF and in fact now operate in such a manner.

This is not interoperability but integratability which is a very unique contribution delivered by the F-35 across the ADF and U.S. militaries fleet of F-35s, USMC, US Navy and USAF.

This is simply not true of a legacy aircraft like the Super Hornet, for in fact that is why the ADF was buying the F-35 in the first place.

And air autonomous systems are not a solution for the three-to-five-year period in and of themselves but might become useful adjuncts as ISR or C2 nodes in a kill web especially as Triton comes on board. There could be accelerated capability to move data from Triton to loyal wingman operations if there is an operational and budgetary space for the USE of autonomous systems prioritized by the government in the three-to-five-year period.

And the work on the Australian approach to agile combat employment is a priority but will be costly up front and require new working relationships between Army and the RAAF as well.

In an interview I did last year with John Blackburn with Air Vice-Marshal Darren Goldie, then the Air Commander of the RAAF, we discussed the challenge of re-focusing the force:

"We don't have the level of knowledge and normative experience we need to generate regarding infrastructure across Western and Northern Australia for the Australian version of agile combat employment."

He contrasted the Australian to the PACAF approach to agility. The USAF in his view was working on how to trim down support staff for air operations and learning how to use multiple bases in the Pacific, some of which they owned and some of which they did not own.

The Australian concept he was highlighting was focused on Australian geography and how the joint force and the infrastructure which could be built — much of it mobile – could allow for dispersed air combat operations.

This meant in his view that "we need to have a clear understanding of the fail and no-fail enablers" for the kind of dispersed operations necessary to enhance the ADF's deterrent capability.

A key element of this is C2. Rather than looking to traditional CAOC battle management, the focus needs as well to focus on C2 in a dispersed or disaggregate way, where the commander knows what is available to them in an area of operations and aggregate those forces into an integrated combat element operating as a distributed entity.

Goldie commented: "We are developing concepts about how we will do command and control on a more geographic basis. This builds on our history with Darwin and Tindal to a certain extent, although technology has widened that scale to be a truly continental distributed control concept.

"We already a familiar with how an air asset like the Wedgetail can take over the C2 of an air battle when communications are cut to the CAOC, but we don't have a great understanding of how that works from a geographic basing perspective. What authorities to move aircraft, people and other assets are vested in local area Commanders that would be resilient to degradation in communications from the theatre commander – or JFACC?

"We need to focus on how we can design our force to manoeuvre effectively using our own territory as the chessboard."

Air Vice-Marshal Goldie underscored that the ability to work with limited resources to generate air combat capability is exercised regularly by the normal activity of 75 Squadron, flying F-35s in Australia's Air Combat Group. This squadron operates from RAAF Base Tindal in the Northern Territory and as Goldie put it: "they have to operate with what they have in a very austere area."

He highlighted a recent exercise which 75 squadron did with their Malaysian partners. The squadron operated their F-35s, and each day practiced operations using a different support structure. One day the operated with a C-27J which carried secure communication, along with HF communications systems and dealing with bandwidth challenges each bearer posed.

Another day they would operate with a ground vehicle packed with support equipment and on another day they would operate without either support capability. The point being the need is to learn to operate in austere support environments and to shape the skill sets to do so.

By learning how to use Australian territory to support agile air operations, and to take those capabilities to partner or allied operational areas, Australia will significantly enhance its deterrent capabilities going forward. This is a key challenge being squarely addressed by the RAAF.[2]

So what can be achieved in the near to midterm along these lines?

In my view, this is a key measure of the credibility of Australian deterrence by denial or whatever other term you might use.

SIX

Multi-Domain Operations in the Maritime Domain: The Significance of Digital Interoperability

The April 11, 2024 Williams Foundation Seminar focused on multi-domain operations in support of maritime security and defence. The progress made as the ADF has been building a fifth generation enabled force needs to be continued in the years ahead.

What is at stake is building an effective kill web enabled force which is built on a digital integration effort to allow the ADF to get best results from its deployed force in the operating area of significance.

We have just seen a real-world example of what this means as the Iran attack on Israel was deflected by a kill web force of sensors and shooters spread across a coalition in support of the defence of Israel.

I highlighted this future in a piece I wrote in 2012 and published in *The Proceedings* entitled the long reach of Aegis. That piece was focused on how F-35 integration with Aegis would yield significant results in defense capabilities.[1]

And when I visited the HMAS Hobart in Sydney Harbour, I was reminded how important a common combat system is for integration across a coalition and one's ability to shape digital integration across a multi-domain force.

After a visit to HMAS Hobart in 2018, this is what I wrote:

The ship introduces a new level of combat capability into the Royal Australian Navy in which the ship's reach is significantly greater than any previous ship operational in the Aussie fleet because of its Aegis Combat system.

It is a key building block in shaping an integrated air-sea task force navy in that the capabilities onboard the ship can contribute to an integrated C2, ISR and strike grid in which the evolving capabilities of the ADF can cover a wider area of operation in the waters surrounding Australia or in service of missions further abroad.

As Rear Admiral Mayer noted during an interview I conducted with him while he was Commander of the Australian fleet:[2]

"We are joint by necessity.

"Unlike the US Navy, we do not have our own air force or our own army. Joint is not a theological choice, it's an operational necessity."

What clearly this means is that the future of the Hobart class is working ways to operate in an integrated battlespace with land-based RAAF F-35s, Tritons and P-8s among other air assets.

Their future is not protecting the carrier battle group, as the Aussies have no carrier.

Rather, their future is to provide air defence for accompanying ships in addition to land forces and infrastructure in coastal areas, and for self-protection against missiles and aircraft.

The skill sets being learned to operate the ship, notably the workflow on board the ship, in terms of the use of data, ISR and C2 systems, working situational awareness throughout the work stations onboard the ship, are foundational for other ships coming to the fleet.

With the coming of the HMAS Brisbane, the HMAS Hobart will no longer be a single ship but the lead into a class of ships.

And with the Australian decision with regard to its new frigates which will leverage the Aegis combat system capability as well, the HMAS Hobart has become the lead into a whole new approach to how the Australian fleet will shape its combat networks as well.[3]

The importance of continuing to build integratability across the fleet was emphasized at the seminar by Liam Catterson in his presentation. He is a former Royal Australian Naval officer who served on the Hobart and operated the Aegis combat system. He is now with Lockheed Martin Australia.

In his presentation he highlighted the significance of the Aegis Combat System for fleet and ADF integrability with the U.S. Navy and Australia's other core maritime allies in the region, Japan and South Korea, all of whom operate Aegis and F-35s.

Catterson underscored the following:

It is important to note that the current fleet of three Hobart-class DDGs are interoperable with the Aegis equipped platforms of the USN, and other Aegis equipped coalition partners, such as the Japanese Maritime Self-Defence Force and the Republic of Korean Navy.

This point was best illustrated through the first operational deployments within the Indo-pacific supporting 7th fleet activities, becoming a integral platform in the INDOPACOM theatre as opposed to previous deployments.

This can be attributed in part to Aegis being as much a fighting philosophy as it is a Combat Management System, melding the concepts of a layered defensive posture, through depth of fire, sensor optimisation, autonomy and integrated fire control through Cooperative Engagement Capability.

Interoperability in the Defence of Australia

- RAN to operate in support of protecting Sea lines of Communication and in support of Anti-Access/Area Denial (A2/AD) mission tasking
- The multi-domain aperture of the maritime mission set increases complexity:
    - Integrated Air and Missile Defence (IAMD), Strike and Anti-Submarine Warfare (ASW) conducted concurrently
    - support from space and cyber, utilising surface, sub-surface and air assets to close the kill-chain
- To combat the threats posed by the significant naval build-up within the region, the need for enablers for **operate** and **integrate** with surface combatants is critical

Slide from Catterson's presentation to the Sir Richard Williams Seminar 11 April 2024.

Without a CMS, a warship ceases to be that; no longer an instrument of deterrence. Without an interoperable CMS, a warship is a well-informed target and a potential hindrance to the joint force.

This is a critical distinction when considering the acquisition of any future classes of surface combatants. The density of VLS cells in an Aegis destroyer is force projection however it is the Aegis Combat System that makes it a force multiplier.

I had a chance to follow up with Catterson in a meeting at his office in Canberra on 15 April 2024. We discussed the way ahead with the digital backbone of a kill web force and the contribution of the common combat system built around Aegis for the Australian fleet and its integration with those of its allies in the region.

We started by discussing how the Aegis combat system enabled significant interoperability across the allied forces in the Pacific.

Liam Catterson attending the Sir Richard Williams Foundation seminar April 11, 2024.

As Catterson noted: "One of the key things about the Aegis combat system operating across the Indo Pacific is that it provides a strong backbone of interconnectivity and interoperability from Australia all the way through the north through to Japan, and then across the Sea of Japan to the Republic of Korea as well.

"The Aegis combat system provides a common language across the Indo Pacific fleets allowing for the for the fleets to deploy and operate together and to conduct combat operations in a coherent manner."

I then raised a key issue. When one mentions the Aegis combat system at a seminar like we had at Williams, one might think that it is special pleading for a specific company, in this case Lockheed Martin.

But over the years the combat system has changed dramatically and it is clearly the U.S. Navy driving the development with Lockheed a close partners, but it is essentially a U.S. Navy combat system today.

Catterson provided an explanation of this development.

"One of the strengths of Aegis is it was developed by the US Navy, and it has been a strong customer holding corporations to account to deliver what they wanted.

"Lockheed has been fortunate to be in lockstep with the US Navy, but it's the US Navy driving these changes for it allows them to embark on the next generation of an integrated combat system for the fleet.

"This will enable them to operate as a system of systems to allow for interoperability, but also to enable cost effective and rapid roll out of developmental changes."

He closed with this thought which is very relevant to the future development of ADF multi-domain capabilities:

"One of the strengths of the Aegis program is leveraging the operational experience from not only the U.S. but also other Aegis users as well. This allows for upgrading the fleet in a spiral development process. And this allows countries to remain in lockstep with each other. This means that integration costs are spread out over different partner nations in that manner."

SEVEN

Multi-Domain Operations in Australia's Maritime Strategy: The Army, Navy, and Air Force Orient Their Efforts

What really does multi-domain mean from the standpoint of each service in pursuit of an effective maritime strategy?

This is determined in part by how one defines what an effective strategy requires and this determination is shaped by whether you are a land, air or surface or sub-surface force.

A multi-domain focus can blur an essential perspective: in particular operations, who is the supported and supported force in pursuit of what outcome or effect desired in an operation?

The Army Role

The changes being worked by the Australian government have a very significant impact on the Australian Army. Not only is their role focused on the region and operations from the northern areas of the country, but their template for operations is shifting as well. They are becoming a littoral maneuver force in support of operations in the maritime regions and areas north of Australia.

And the USMC rotation to the Northern Territories will be part of shaping that template. It should be noted that the Marines are working their open template for operations throughout the region, and the Australian Army and USMC will almost certainly dovetail operations.

As they template is shaped, it is obvious that funding or new equipment needs to be provided. Some is already in place in terms of providing for longer range strike and for ships to move Army forces within the region.

As a USNI News piece described the changes in an October 2, 2023 article:

The Australian Army is slated to shift its focus to the littorals after announcing last week several major changes, which include the redeploying a sizable portion of soldiers and equipment across the country and optimizing several brigades for littoral and amphibious missions…

The Australian Department of Defence announced these changes in response to the 2023 Defence Strategic Review…The DSR recommended it to be "optimized for littoral operations in our northern land and maritime spaces and provide a long-range strike capability."

Aside from reducing the procurement of infantry fighting vehicles and self-propelled howitzers, some of the top recommendations for the Australian Army were to speed up the procurement and increase the quantity of HIMARS, land-based maritime Strike systems and amphibious vessels.

Last week's announcement highlighted significant changes to the mission sets of the 1st, 7th and 3rd multirole combat brigades, which will become more specialized.

The 1st Brigade will be transformed into a light combat brigade, which will allow it to be "light, agile and quick to deploy in the littoral environment" and "support land-based long-range fires." While Australia has ordered HIMARS, under LAND 4100 Phase 2 the Australian Army is looking to procure a land-based maritime strike capability…

The 7th and 3rd will become motorized and armored combat brigades, respectively. However, like 1st Brigade, the two also will focus on littoral and amphibious operations. To address these littoral missions, brand-new littoral lift groups are also slated to be created and collocated with the brigades in their respective basing locations.

Littoral lift groups will host Army Littoral Manoeuvre Vessels, including both landing craft medium and heavy, which will be procured in Phases 1 and 2 LAND 8710…

At the moment, 2nd Battalion, Royal Australian Regiment, is Australia's premier and only amphibious-focused unit. With the changes announced last week, all three of the Australian Army's active brigades will have either littoral or amphibious focuses.[1]

Brigadier James Davis presenting at the Williams Foundation April 11, 2024.

On the one hand, the Army is to play a role in supporting maritime operations by being able to deliver strike in support of maritime forces. On the other hand, the Army needs to have sufficient size to hold ground in significant areas out to Australia's first island chain in time of conflict, and the Army then would be the supported force.

The USMC unlike the Australian Army has organic lift and long-distance assets such as the Osprey and the F-35B which can support its littoral operations. The Australian Army is a rotorcraft enabled force without the kind of lift which the Osprey and the CH-53K provide the USMC. And the integration of the F-35 into the USMC maneuver element is a key element changing how the rest of the littoral force can operate. Will a similar role occur with Australian F-35s and the Australian Army?

At the seminar, the Army perspective was provided BRIG James

Davis, Director General Future Land Warfare. In his presentation, he underscored to the audience that "the majority of the infrastructure which supports a maritime strategy is on land."

In that sense, littoral maneuver from one land location to another within the littoral maneuver space.

"Ports, airports, sensors, satellite dishes, terrestrial launch and recovery are land-based. For context, Australia has 59,000 kilometers of coast and 50% of our population live within a few kilometers of our coasts. Beyond are shores but within our sovereign area are 8,222 islands and numerous offshore installations."

He underscored that the "DSR described an Army optimized for littoral and archipelagic operations." And here he provided a clear sense of the Army role and perspective: "The littoral is an area for the fusion of cross-domain effects and where land forces can make their greatest contribution to the integrated force."

He argued that the government has therefore shifted resources within Army to work in this domain. "This includes government direction to establish a new long range fires regiment equipped with 36 HIMARS launchers and a littoral group of 18 medium watercraft, pending approval of a second long-range strike regiment, and eight heavy landing craft will be in service from the end of the decade."

A key role for the Australian Army is working in the neighborhood. "In peacetime, army watercraft will operate to provide organic mobility to the integrated force and to work with our partners and allies in the region, building collective understanding capability and offsetting risk, because armies are the largest arm of the militaries in our region."

He then added: "In conflict, we see special and general-purpose forces using these vessels operating in operations below the engagement threshold. They will be able to enable joint and integrated C4 by getting communications nodes and relays to the right places, and getting sensors, weapons and influence to where they can exert domain control largely in the maritime domain.

"This includes maritime strike systems with ranges of hundreds of kilometers. The value of these systems will be the difficulty of detecting or engaging them."

He noted that "land forces will also support all domain targeting."

And control of territory within the littoral region is crucial as well. He underscored that "at times, more robust application of land power will be needed to maintain or gain control of specific terrain, such as offshore islands. These outposts have always been critical in maritime strategy."

He provided a good description of the new template. He highlighted some of the initial investments to make the template real but there are significant changes in aviation as well as watercraft, including in autonomous systems to be made and paid for in the years ahead.

The Air Force Role

In the presentation of Air Commodore Mick Durant, Director General Strategy and Planning—Air Force, the role of the RAAF in maritime operations was highlighted.

Given that the RAAF provides the air element for the Royal Australian Navy this is somewhat equivalent to a discussion of how the U.S. Navy's air arm works with the fleet and then with the USAF, but it is different because the integration of the RAAF and the RAN is a key element of the operational realities of the ADF.

Their integration already is multi-domain so what is necessary is to sort out how the addition of the SSN's alter this and how the new fleet elements will work to reinforce or disrupt integration already created by the 2017 government focus on the Aegis combat system being the digital backbone of the fleet which has enabled deeper RAN and RAAF integration, and in fact such digital integration is crucial to shaping multi-domain operations.

As he commenced his presentation Durant highlighted the operational challenge: "From an airpower viewpoint, we will operate at distance from our home bases from austere and remote locations across our north and operate deep into the Indian Ocean and Pacific Ocean and the surrounding areas."

And closing kill chains across a vast region highlights the need to integrate sensors to deliver to weapons effects across the combat chessboard.

Air Commodore Mick Durant presenting at the Williams Foundation Seminar April 11, 2024.

In effect, the RAAF provides both sensors and shooters in the maritime areas of operation, and sensors to enable the fleet and its targeting efforts as well. Notably, the coming of Triton is symptomatic of these integration efforts whereby targeting data is generated outside of the weapons engagement zone and transmitted to other sensors and to shooter in the engagement zone.

As Durant underscored: "Our potential adversaries will also be highly adaptive, and we are seeking to do the same. This also draws a requirement for the integrative force to think more deeply about building resiliency, as well as managing its own signature. All of this is underpinned by the air force intelligence and enterprise targeting capabilities.

"With the introduction of new platforms such as the F 35, the P-8, the Triton, and the Peregrine, the complexity and volume of intelligence data has and will continue to increase exponentially. Defense intelligence capabilities will need to embrace automation and edge processing to accelerate Association, correlation, and fusion of data within high capacity resilient and redundant networks."

He emphasized the air agility basing issue in which the ISR and C2 systems needed to embrace a dispersed force operating from various locations on Australian territory. The intersection of a kill web C4ISR system with force distribution is a crucial way ahead for the RAAF.

Air Commodore Durant put it this way: "The key airpower principle of centralized control may prove to be transient. However, distributed and decentralized clusters will be able to generate both deliberate and dynamic air effects. To contextualize this through an integrated air and missile defense lens, we will never have enough exquisite interceptors to interdict all threats and to protect all key nodes.

"This not only reinforces the criticality of passive protection measures of camouflage. deception and hardening, but it also underscores the need for new approaches to create distributed mass. A more asymmetric force mix that includes uncrewed and autonomous systems, to complement the force in being is how a small to medium size Air Force might generate greater mass lethality and survivability into its air combat system."

He highlighted progress underway, such as the creation of regional air force development teams across northern Australia to examine how to enhance force posture options.

Air Commodore Durant provided an example of progress evidenced in the last Talisman Sabre exercise. "Last year's exercise provided a huge opportunity for Air Force and Navy to integrate the ADF its most potent air defence maritime capabilities with HMS Sydney and Hobart integrating their air and missile defense capabilities with the Wedgetail aircraft. Combined with an equally potent force of P-8s, Growlers and F 35 stealth fighter aircraft, the participants exercised against an equally formidable threat environment where emerging capability capabilities were trialed.

"Our emerging agile control teams established remote command and control linkages for the joint force demonstrating an ability to pass data using multiple discrete nodes providing resilience to a connectivity matrix in a denied environment.

"Additionally, our maritime strike platforms exercised and tested their ability to find, fix and target discrete maritime assets at tactically significant ranges. Against an equally challenging threat environment, the joint force exercised their ability to introduce new tactics and procedures throughout the exercise to counter emerging threat capabilities."

Air Commodore Durant provided another example of the way ahead with regard to operational innovation evidenced in the coalition exercise Cope North.

This is how he put it: "The focus of the exercise was to stress, validate and improve national and trilateral agile combat employment capabilities. Commencing in Guam, the exercise saw the activation and operation of United States Air Force and Japanese self-defence force and our own RAAF assets from the main operating base to six Island forward operating bases. Over the course of three weeks. Air Force representatives from Air Command innovation and Jericho joined the exercise to work with the USAF combined rapid capability development team.

"This team was focused on solving critical operational problem sets as they arose in theater, or as a response to adverse reaction with the ability to rapidly deploy the solution to the frontline in order to maintain the competitive advantage of the coalition effect."

He concluded: "The ability for airpower to deliver impactful projection within our maritime approaches requires a combination of effective defence, combined with a series of highly integrated multi-domain offensive counters as part of the integrated force and in conjunction with allies and partners. This is how airpower will deliver a strategy of denial in our key maritime approaches."

The Navy Role

The perspective of the Royal Australian Navy was presented by Rear Admiral Stephen Hughes, Head of Navy Capability. Hughes presented at the last seminar as well and there he provided a number of insights.

At that seminar, RADM Stephen Hughes, Head Navy Capability, underscored that when focusing on the maritime domain, one is inherently focused on multi-domain strike. The maritime warfighting domain is shaped by strike whether coming from land, surface, subsurface or air domains, as well the cyber and space domains.

Rear Admiral Stephen Hughes presenting at the Williams Foundation Seminar April 11, 2024.

"To attain long range strike capabilities allows us to move from a strategy of defense to a strategy of deterrence through denial which signifies a national shift that aims to hold an adversary at risk a greater range raising a question in the adversary's mind about whether they want to attempt to act against us.

"So what does the maritime force bring to the fight?

"A maritime force is able to be agile, mobile expeditionary scalable, sustainable, versatile, networked, and lethal. Maritime force provides critical advantages through their ability to use the oceans to maneuver and hide in the case of submarines, and the airspace and the space above that domain. Maritime force combines distributed fleet operations, and mobile expeditionary forces with sea control and sea denial capabilities.

"However, a maritime force does not compete, deter, or fight alone. The maritime force is an integral part of the joint force and works closely with allies and partners to bring to bear maritime effects. Controlling the seas enables the maritime force to project power in support of Joint Force efforts. surging into the theater of operations, where adversaries must cross open water. Sea denial deprives them the initiative prevents them from achieving their objectives.

"Maritime force controls or denies the seas by destroying an adversary's fleet or their associated air support. And in in the modern battle space even extends into space. It can contain it in areas that prevents meaningful operations prohibited from leaving port by controlling sea lines of communication. Maritime forces capable of controlling critical choke points enable joint forces to impose military and economic costs on the adversary."

He also added comments with regard to the innovation which Navy is working to enhance multi-domain strike. "The future of our strike capability needs to include the use of uncrewed systems. Navy is working with industry in exploring solutions through the autonomous warrior series of experimental exercises. And such systems will have the ability to strike deep against an adversary by deploying mines and other guided weapons by using sovereign Australian capabilities."

As he underscored: "The defense strategic review has placed a premium on accelerating lethality for deterrence and impactful projection,"

He cited the examples not only of acquiring TLAMS but the development of greater maritime strike capabilities. against maritime forces, whether from an F-35 or from anti-ship missile capabilities.

At this seminar, he added an update with regard to what the Australian government has focused on in its strategic shift, namely, a focused force on the region which will eventually include nuclear attack submarines and new surface ships as announced in the fleet review recently announced.

As the government announced on February 20, 2024, the intention was to expand the surface fleet.

Today, the Albanese Government has released its blueprint for a larger and more lethal surface combatant fleet for the Royal Australian Navy, more than doubling the size of the surface combatant fleet under the former government's plan.

This follows the Government's careful consideration of the recommendations of the independent analysis of the surface combatant fleet, commissioned in response to the Defence Strategic Review.

Our strategic circumstances require a larger and more lethal surface combatant fleet, complemented by a conventionally-armed, nuclear-powered submarine fleet.

Navy's future fleet will be integral to ensure the safety and security of our sea lines of communication and maritime trade, through operations in our immediate region. This fleet will constitute the largest number of surface combatants since WWII.

The independent analysis of Navy's surface combatant fleet lamented the current surface combatant fleet was the oldest fleet Navy has operated in its history, and emphasised the need for immediate action to boost Navy's air defence, long-range strike, presence and anti-submarine warfare capabilities.

In line with independent analysis' recommendations, Navy's future surface combatant fleet will comprise:

- *26 major surface combatants consisting of:*
- *Three Hobart class air warfare destroyers with upgraded air defence and strike capabilities*
- *Six Hunter class frigates to boost Navy's undersea warfare and strike capabilities*
- *11 new general purpose frigates that will provide maritime and land strike, air defence and escort capabilities*

- *Six new Large Optionally Crewed Surface Vessels (LOSVs) that will significantly increase Navy's long-range strike capacity*
- *Six remaining Anzac class frigates with the two oldest ships to be decommissioned as per their planned service life.*

The Government has also accepted the independent analysis' recommendations to have:

25 minor war vessels to contribute to civil maritime security operations, which includes six Offshore Patrol Vessels (OPVs).

The Hunter class frigates will be built at the Osborne shipyard in South Australia, and will be followed by the replacement of the Hobart class destroyer. The Hobart destroyers will be upgraded at Osborne with the latest US Navy Aegis combat system.

The new general purpose frigate will be accelerated to replace the Anzac class frigates, meaning the Transition Capability Assurance (TransCAP) upgrades are no longer required. These new general purpose frigates will be modern, capable and more lethal, requiring smaller crews than the Anzac.

Consolidation of the Henderson precinct is currently underway, as recommended by the Defence Strategic Review. Successful and timely consolidation will enable eight new general purpose frigates to be built at the Henderson precinct, and will also enable a pathway to build six new Large Optionally Crewed Surface Vessels in Western Australia.

The Albanese Government is committed to continuous naval shipbuilding in Australia and the design of Navy's future fleet will provide a stable and ongoing pipeline of work to the 2040s and beyond.

Although the review described projected fleet size, the crucial question is how the subsurface and surface fleet, crewed and uncrewed, are integrated in or integratable within a kill web force.

It is about the effects created through such a force rather than simply having a ship building program in my view. In fact when a colleague and I were working for a senior U.S. Navy Admiral, we began to work on a project that he thought was long overdue, namely, replacing the 30 year shipbuilding program with a very different measure, namely a 30 year Navy capability plan. The words are significant here.

This is in my view one of the challenges is using a phrase like multi-domain as it may obscure what the real objective is, namely, to project power and to have effective mobile defence of your forces and nation to deliver the desired combat and strategic effects.

The future of the Royal Australian Navy rests within this matrix, and as Vice Admiral Barrett has argued in my recent interview with him, that ramping up capability in the three-to-five-year period rests ultimately on the ability to shape operational space to use autonomous systems,

As he stated: "The surface combatant review took an eye to considering autonomous systems but considered them a generation away. But the reality is that we are already down the autonomous systems path now.

"It is wrong simply to focus on long range prospects for autonomous systems not yet here, such as platforms which could potentially carry a large number of weapons cells, rather than on the systems that are already here. The current systems can deliver significant ISR capability for example, and we need to integrate these systems into the operating force."

The other key consideration is the integration of the combat systems in the surface and subsurface fleet in a way that allows for the kind of integration mentioned earlier in the Air Force perspective, namely the air warfare destroyer's integration of air force combat systems.

In 2017, the Australian government took a key decision which in my view is crucial to maintain.

This was the October 3, 2017 announcement:

The new approach for combat management systems will ensure our Navy's future ships are fitted out to protect Australia in the decades ahead.

Under the plan, the combat management system for Australia's fleet of nine Future Frigates will be provided by the Aegis Combat Management System, together with an Australian tactical interface, which will be developed by SAAB Australia.

This decision will maximise the Future Frigate's air warfare capabilities, enabling these ships to engage threat missiles at long range, which is vital given rogue states are developing missiles with advanced range and speed.

The Future Frigates will be operating in a complex and growing threat envi-

ronment. By bringing together the proven Aegis system, with a cutting edge Australian tactical interface developed by SAAB Australia, our Future Frigates will have the best capability to defeat future threats above and below the surface, while also ensuring we maintain sovereign control of key technologies, such as the Australian designed and built CEA phased array radar.

In the past, Defence has taken the tendered combat management systems individually, which has meant that the Navy has operated numerous systems at the same time. This has not allowed defence industry to strategically invest for the long-term and has also increased the cost of training, maintenance and repair.

Under the Turnbull Government's new strategic enterprise approach, the Government has now mandated that where the high-end warfighting capabilities of the Aegis system are not required, a SAAB Australia developed combat management system will be used on all of Australia's future ship projects.

This includes mandating a SAAB Australia combat management system on the upcoming Offshore Patrol Vessels, which will be built in Australia from 2018, and an Australian tactical interface developed by SAAB Australia for the Hobart class Air Warfare Destroyers when their Aegis combat management system is upgraded in the future, consistent with the 2016 Defence White Paper.

Further, it guarantees the development of a long-term sustainable Australian Combat Management System industry, which is integral to the implementation of the Government's Naval Shipbuilding Plan.[2]

Frankly, nothing has changed for the necessity of such an approach. And in fact arguing for a multi-domain integrated Navy only underscores its necessity.

EIGHT

Key Dimensions Required for a Successful Multi-Domain Approach

Several specific aspects of defence capability were highlighted in the 11 April 2024 seminar as key dimensions of a multi-domain force.

The first was joint C2; the second was the role of space in enabling such an effort; the third was the pacing function for such a force, namely, logistics and sustainment.

Shaping C2 for the ADF and Coalition Forces: The Perspective of Air Vice Marshal Mike Kitcher

When I was in Australia in September of last year, I had a chance to talk to the then Deputy Chief Joint Operations (DCJOPS), Air Vice-Marshal Mike Kitcher. I discussed that interview in the second chapter of thus book. At the April 11, 2024 Williams Foundation seminar, Kitcher focused on the C2 aspect involved in the changes we talked about last year.

Distributed C2 for the ADF and C2 directing coalition operations are critical challenges to be met as the ADF adapts to the operating the "focused force" the government has mandated.

Air Vice-Marshal Kitcher presenting at the Williams Foundation Seminar April 11, 2024

At the seminar, Kitcher underscored that "we are focused on building a headquarters that's capable of planning, executing, managing regional operations, from competition through crisis to conflict."

He underscored that the ADF was working on a model different from the American model of the combatant commander.

"The size and scale of the personnel involved in a U.S. Combatant Command compared to the ADF is very different. When we add component commands to a joint operation, we need to have a need to consider the numbers of people that we have available in our component command model. And we need to cut our cloth to the numbers that exist realistically."

I would personally add observing American command structures that they have generally been very large, and a key change underway is to shift to distributed C2 which is forcing changes in terms of the size of strategic or theater level command.

And not surprisingly, ADF work in this area has an influence on those military commanders who are actually working in innovative ways with regard to C2 innovations,

When Vice Admiral Lewis became commander of the Second Fleet, he focused specifically on how to lean out command elements and empower distributed forces to execute mission command.

This is a subject which we discussed in some detail in our various visits to the Norfolk-based command.

He provided an example of the changes being worked by JOC as seen in the last Talisman Sabre exercise.

And we discussed that further in a meeting later in the month. At that meeting, he discussed with me further the changes in the operational command and control approach.

He argued that "at the seminar, I discussed a component C2 model in which six components were being blended in, namely six components, space and cyber as well as the more traditional air, land, maritime and special forces.

At Talisman Sabre we introduced we shaped a logistics coordination command for the entire coalition effort for nearly 30,000 people involved in the exercise.

He assessed the state of the art to date as follows: "We are working to understand the supported and supporting commander roles within the components, and which components are relatively mature, which components have to mature, which components are well versed and operating as components and which components are working hard to design and execute their component functions.

"With regard to our maritime environment where we find ourselves in our region, all of the six components could be the supported commander for particular periods in particular events. But broadly speaking, the air and maritime component is the most logical components to lead in as the supported command with the other components supporting the air and or maritime component across the spectrum of operations.

"And the key relationship then becomes that between the component command and the JOC at the operational level on how to successfully integrate those two functions."

We then turned to the recent Talisman Sabre exercise experience.

According to Kitcher: "For the first time, we had a single leader of U.S. forces at the Corps level working the U.S. engagement. And each coalition nation had that level of leadership deemed appropriate for the size and scale of their involvement in the exercise.

"That is a model we will continue with. Within JOC we

embrace that leadership model, and we embrace as well the engagement various different government departments such as the Australian Federal Police and their embedded liaison officers in JOC as well."

In short, the kind of impactful presence which Australia is building in the region, C2 is a key element for creation of enhanced capability in the defence of Australia.

How to Enhance Space's Contribution to Australian Multi-Domain Operations in Support of Maritime Operations

Australia has more demands than it has budget. It has more challenges than it has forces.

So how to maximize the effect of what capabilities Australia has going forward?

How to maximize the economy of force to deal with persistent and expanding threats?

Part of the answer is to focus on enhanced leveraging of what can be generated from the space domain. The strategic shift in what space delivers in the past few years is driven by the arrival of constellations of LEO satellites which provide reach and coverage for data of various sorts, including of course communications which is historically unprecedented.

The challenge for governments is how to leverage such constellations and create government organizations, strategy, and support to do so. Australia is no different but capabilities and resources more limited than they have been for the U.S. or several of its other allies.

Leveraging the commercial sector and its innovations then is even more critical for a modest space player like Australia. This would require a smart sovereignty effort but focuses effort and investments from its government.

At the Williams Foundation seminar on April 11, 2014, Nick Miller, Senior Director, Space Solutions & Strategy, Optus Satellite & Space Systems, provided his perspective on a way ahead. He started by characterizing the stakes of the game as follows:

"Without space, there would be no effective mult-idomain capability. It is a critical enabler. Space once deemed the final frontier is now seen as a potential battleground. In this domain, where the stakes are as high as the orbits, the mastery of space technology is not merely an option, it is vital for our national security."

He told the audience that Optus as Australia's leading satellite owner and operator "transmits via its fleet 27 gigabits of data every minute of every day."

Nick Miller speaking at the April 11, 2024 Williams Foundation Seminar.

Miller noted that "with a robust enterprise satellite network, Australia can improve its ability to conduct comprehensive ISR operations and as a nation with vast maritime domains and extensive borders, satellite data is crucial for maritime operations and to protect vital sea lines of communications for Australia as well as enhancing our Border Force."

Several speakers at the seminar highlighted the importance of undersea cables for transmitting data into and out of Australia in support of the Australian way of life.

Miller argued that satellite networks provide redundancy to ensure flow of data in times of disruption. "As part of readiness preparation, a comprehensive satellite network would support critical and effective command and control capabilities for the ADF and support a whole of government approach in situations such as natural disasters, search and rescue operations, and national emergencies where the subsea cables may have been compromised."

Much of his presentation argued for a robust commercial and government working relationship including government funding to deal with the increasing challenge of protecting operational space assets. Not only do adversarial powers plan and practice space denial, but the impact of space debris on operational constellations is significant.

Such joint efforts between government and the commercial sector in space are crucial to ensure that Australia has the skilled workforce necessary to support broader space efforts, and to have the required expertise in times of crisis.

Miller underscored: "Luckily in Australia, we have some personnel with nearly 40 years of experience and this is something we're offering the defense sector and can be a foundation from which to expand the skill base."

But to have the kind of effective public-private partnership which Australia needs requires an innovative acquisition process.

Miller opined: "There needs to be a capability to fast track satellite technologies. Flexible contracting models are needed to provide incentives for Australian companies and research institutions to innovate more rapidly in satellite technology.

"There is a need to bolster local industry and academia grants, tax incentives and government contracts awarded to homegrown enterprises encouraging domestic innovation and reducing dependence on overseas providers.

"Optus itself is looking into how a partnership with universities, startups and government could deliver a new and unique novel LEO capability."

Professor Andrew Carr recently underscored that strategy is not about writing documents with lofty words and concepts – it is about finding ways to identify and address core problems with realistic solutions.

As he wrote recently: 'Strategy as problem solving' shifts the emphasis from declaring our principles to diagnosing our problems. The key work of Australian strategists in the years to come will be twofold: to identify which problems are most important, based on their significance, the likelihood of harm and how we might resolve them; and to interrogate their dynamics, understanding why they're so hard and where leverage points may be found to seek better patterns of order."[1]

Miller presented a thoughtful way to proceed with strategy understood in Carr's terms.

Logistics and Sustainment for an Evolving Defence of Australia Strategy

There is no more daunting challenge facing a credible Australian defence strategy in it is region than that of logistics and sustainment. Australia is so dependent on imports of supplies, and overseas production of military equipment that the nation is very exposed to its logistics and sustainment shortfalls.

And when looks at the wider world of its allies, the picture is not brilliant. The war in Ukraine and the challenge for Europe and the United States to provide basic supplies has been daunting.

Bluntly put, the democracies have moved from their industrial base and have not built defense in depth. The only country in the West that remained committed to national mobilization was Finland, where I conducted several interviews during a past visit precisely on how they have addressed how they have built in mobilization from the ground up.

If Australia is to have a credible logistics and sustainment foundation, investments need to be made in supplies and stockpiles, in the building of industrial base – probably through joint efforts with allies – accepting the need for industrialization in key areas, an energy policy that leverages their natural supplies and capabilities, and working with South Korea, Japan and the United States on an innovative way to enhance Australia's potential role as a strategic bastion in the Pacific based on enhanced supplies, support structures and production capabilities in Australia which allies invest in as well.

But whether or not Australia can achieve this is a major challenge which will require investments significantly beyond what the government is contemplating and an engagement with industry that requires a major shift in how the defense industrial base is built, sustained, and how the ADF can work much more directly in the development of evolving capabilities, such as autonomous weapons.

MAJGEN Jason Walk, Commander Joint Logistics, addressing the April 11, 2024 Williams Foundation Seminar.

At the April 11, 2024 Williams Foundation seminar, MAJGEN Jason Walk, Commander Joint Logistics, provided an overview of how the logistics challenge is being defined and focused on in the wake of the Defence Strategic Review.

This is how characterized the change: "The DSR directed that the defence logistics network be adequately resourced to deliver persistent support and sustainment for operations. This considered by itself is a step change in defence capability and capacity, demanding first, that defence first confirm its logistics gaps before embarking upon the most substantial investment in the defence logistics network, arguably since World War Two."

After discussing the need for robust cyber defence to protect the network, and shaping space based capabilities to support such a network, he then turned to the question of the near-term focus.

"So what are the problems we're trying to address within the defence logistics network?

"The log network underpins defence's force posture, ensuring the right stuff gets to the right location at the right time. Accordingly, the defense strategic review required the ADF to develop a Northern Australian network of bases to provide a platform of logistic support for denial and deterrence. To address this, defence's ambition for the defence logistics network can be summarized as making a more a more agile, effective, integrated and resilient network."

I have spent a great deal of time with the various logistics commands in the United States and seen over the last thirty years significant change in multi-modal logistics.

But the ability of the U.S. to deploy military power relies very heavily upon commercial systems which will be difficult to depend on in times of crisis, more limited air lift and tanker capacity than would be needed for the USAF alone, let alone for the U.S. Army, and a Military Sealift Command whose capability is limited by the decline of the US merchant marine, for MSC is operated by mariners not the U.S. Navy. The Navy is buying Ospreys because of the limited lift capabilities available to the Navy.

So what then about Australia?

Will we see an upsurge of the means to provide for the support a distributed and more mobile force will need.

MAJGEN Jason Walk addressed this challenge as follows:

"First of all, to an agile and multi modal logistics. A key logistics problem that we face is the paucity of strategic maritime lift capabilities to enable the projection and sustainment of forces.

"The solution is to build a diverse multi modal logistics network that leverages a mix of transportation capabilities across land, sea and air. This will allow the agility to rapidly reorganize reprioritize and adapt the delivery of logistics effects in response to changing requirements or threats.

"Leveraging industry support in areas of reduced threat will enable the focused application of limited ADF strategic lift assets.

"One initiative under consideration by government is the establishment of a maritime strategic fleet. Importantly, this reflects our support and sustainment of military capability will require a whole of government indeed a whole of nation endeavor. Effective logistics with increased stocks, the ADF in Australia at large lacks sufficient stocks of critical commodities, like explosive ordnance fuel to sustain operations, especially in our northern regions.

"The solution is to invest in depth and redundancy to build up strategic reserves and material stocks to meet the demands of the integrated force.

"In conflict, these critical supplies must be replenished and forward positioned to optimize operational availability and freedom of action of our deployed forces. This is a focus of a number of defense projects moving forward and we are establishing closer linkage with other government agencies and industry.

"The National Fuel Council which had its inaugural meeting early last year or mid last year is an example of that. The integration of logistics across all domains and coalition's is equally important, logistics interoperability that can support an integrated force and operations in coalition with allies is critical.

"The solution is to design an integrated logistics network that reduces friction and complexity. The defence logistics network seeks to minimize organizational seams across the defence enterprise and reinforce interfaces with industry, whole of government, allies and partners."

The speaker provided a good description of the challenge.

But frankly, this is a daunting one, in which phases needed to be shaped and credibly funded. This will not be critical just for the ADF but for an credible cooperation policy in which South Korea, Japan and the United States would participate in effectively.

And let me be blunt: this would have to be addressed in real terms by investments and policy changes by those allies as well. This is not about a focus simply on AUKUS – this is about building a credible and real arsenal of democracy in our time.

AUKUS can too easily used as a Rorschach image where one can see what one wants. It is not an end in itself. If meaningful, it is a gateway to solving a strategic challenge such as that discussed by the speaker.

NINE

Cognitive and Information War and the "Gray Zone"

An aspect of modern Western strategic thinking has been a focus on gray zone conflict.

This is an area I have always found confusing.

In a world which I would characterize as one of the rise of multi-polar authoritarian movements and states, their constant conflict efforts are indeed been in the gray zone punctuated with direct periods of violence against the West and its legacy of a "rules-based order."

But as this is going on, it would be difficult not to factor in the domestic conflicts in both the UK and the United States which affects the AUKUS partners of Australia. So how well is Australia doing in the gray zone or information or cognitive warfare areas?

The is a major aspect affecting any credible strategy involving a " whole of government" strategy or a whole of society effort to deal with threats in the region.

The West over the past few years has done considerably better in the cyber-war domain, but given the penetration of authoritarian movements and states within our social networks, and the extensive disruption in the West with regard to migration, I do not think we can make the same judgement with regard to information or cognitive war.

At the Williams Foundation Seminar on April 11, 2024, the subject of information war was addressed by Major General Anna Duncan, Commander Cyber Command. Her talk highlighted the importance of gaining information advantage in conflict.

Major General Anna Duncan, Commander Cyber Command, presents at the Williams Foundation Seminar, April 11, 2024.

She started with this definition: "What is information advantage? From a military perspective, information advantage, ideally occurs through the integration and through the use of the moral and information informational elements of fighting power. We would seek to gain an information advantage over an opponent by targeting their understanding and thus degrade their will to fight."

She cautioned that was not new in warfare but clearly what is new is the nature of information networks in liberal democracies and how conflict has escalated within these societies by the emergence of tribal clubs which operate within social media which has challenged the ability of democracies to shape consensus.

When I attended a UNESCO event in Barcelona in 1996 which

focused on the new information society, I highlighted this danger associated with an internet society. But the extent to which the tribes have grown to disaggregate democracies was certainly not my thought at the time.

The point is important – precisely in the 1990s when many were trumping the global ascendency of democracies, we were building tools which would in fact undercut that ascendency.

Gray zone conflict in my view goes hand in hand with information warfare. Western militaries are building more flexible militaries which can operate as a more distributed force but we have not seen the adaptation of the political class to how in fact confront adversaries in the gray zone effectively nor how to use penetration of authoritarian societies or movements to our advantage.

Duncan provided a professional treatment of how the ADF is working through how in conflict to gain an information advantage over adversarial forces. A military officer dealing with cyber and information warfare scopes the focus on information advantage over adversarial forces in a conflict.

This is obviously crucial, but the actual conduct of information war occurs every time an authoritarian government or movement defines the perceived geopolitical reality inside Western societies.

A murderous organization like Hamas defines the ideas for a protest at my former school, Columbia University, due to their information war capabilities.

I would close by including an article I published in December 2021 which underscored gray zone conflict which I also thinks expand the notion of what is entailed in the kind of information war which the West is not very good at engaging in.

Western analysts have coined phrases like hybrid war and gray zones as a way to describe peer conflict below the level of general armed conflict.

But such language creates a cottage industry of think tank analysts, rather than accurately portraying the international security environment.

Peer conflict notably between the liberal democracies and the 21st century authoritarian powers is conflict over global dominance and management. It is not about managing the global commons; it is about whose rules dominate and apply.

Rather than being hybrid or gray, these conflicts, like most grand strategy since Napoleon, are much more about "non war" than they are about war. They shape the rules of the game to give one side usable advantage. They exploit the risk of moving to a higher intensity of confrontation.

Russia is doing this right now in Ukraine. China, likewise, is doing it in the South China Sea and in the Sea of Japan. It's critical to understand this point, and terms like gray zone operations and hybrid war don't capture the challenge of escalation control.

There are two games being played. One game is over the immediate contentions of the major powers. Ukraine and Taiwan must be protected from attack.

But the second game is just as important, it asks what limits should be crossed to manipulate the risk of going to a higher intensity of competition?

In the Cold War these limits defined the "system dynamics" of the competition. Shaping them was important, because they were the foundation for winning a war that might erupt, or toward stabilizing a competition in a way that gave advantage to one side or the other.

Seen this way Korea, Vietnam, Berlin, etc. were about winning those local wars. But they were more importantly about shaping the global competition between the United States and the Soviet Union.

Quite elaborate rules were worked out for this. It took substantial time during the evolution of the Cold War (to make sure that it was indeed was a cold war from a global conflagration point of view) for this learning curve to develop. Limited wars, like Korea, produced know how about escalation control and dominance.

The problem today is that we are only at the earliest parts of this learning curve for our age. We're in a long term competition with authoritarian powers, but it's like it was 1949 in terms of our know how for managing this rivalry to our advantage. The problem isn't simply to defend Ukraine and Taiwan; it's to do it in such a way that doesn't lead to crazy escalations or that doesn't scare the daylights at of our allies.

Taiwan and Ukraine are not sideshows to global conflict; they are the early test cases of competition in a second nuclear age.

Recently, I discussed the question of how best to describe the terminology to describe peer conflict with my colleague Dr. Paul Bracken the author of The Second Nuclear Age.

According to Bracken, it is preferable to use the term "limited war" to describe the nature of conflict between the authoritarian powers and the liberal democracies. "A term was invented in the Cold War which is also quite useful to analyze the contemporary situation, namely, limited war. This term referred to conflict at lower levels and sub-crisis maneuvering. And that is what is going or today in cyber and outer space, to use two examples. But it also applied to higher levels of conflict like limited nuclear war."

"The notion of limited war focuses escalation as a strategy. What is the difference between limited and controlled war?

"That's a really important question with enormous implications for command and control. Today, for example, limits are determined in a decision making process whereby the Pentagon goes to the White House and says we'd like to do this operation. The White says yes or no.

"Left out of this is any discussion of building a command and control system for controlled war. This means keeping war controlled even if things go wrong — as they always do. Without an emphasis on controlled war, and not just limited war, I would estimate that the United States will be highly risk averse, that is, the fear of an escalation spiral will drive the United States toward inaction.

"Look at the Ukraine. The first U.S. reaction to the Russian buildup was to immediately take military options off the table. The White House refocused its strategy on financial sanctions instead. It looked as if the United States was desperately searching for ways not to use force. Soft power, gray zone operations, the weaponization of finance — these are clearly important and I think we should use them.

"But they look like a frantic attempt to any use of force, like British foreign policy in the 1930s.

"Our language shapes our strategy. An image of war that blows up, that's unlimited, or that you've declined to fight because of your fear that it would become so is where we are. In academic studies and think tanks the focus is overwhelmingly on "1914" spirals, accidental war, entanglement, and inadvertent escalation.

"If it's going to be controlled or limited, how are you defining that it is limited? Is it limited by geography? Is it limited by the intensity of operations? Is it limited by the additional political issues that you will bring into the dispute?

"These are never specified in discussions that I see of hybrid or gray zone

warfare. To use a very sensitive example. In a Taiwan scenario, will the United States Navy and Air Force be allowed to strike targets in China?

I see a real danger that this isn't being thought through. If we think it through only in a crisis we're likely to find a lot of surprises in how the White House and Joint Chiefs of Staff see things differently.

These expressions – hybrid war and gray zone conflict – are treated as if they self evident in term of their meaning. Yet they are part of a larger chain of activities and events.

We use the term peer competitor but that is a bit confusing as well as these authoritarian regimes do not have the same ethical constraints or objectives as do liberal democratic regimes. This core cultural, political and ideological conflict who might well escalate a conflict beyond the terms of what we might wish to fight actually.

And that really is the point – escalate and the liberal democracies withdraw and redefine to their disadvantage what the authoritarian powers wish to do.

Bracken noted: "That's a good distinction too, because it brings in the fact that for 20 years we've been fighting an enemy in the Middle East who really can't strike back at the United States or Europe other than with low-level terrorist actions. That will not be the case with Russia, China, and others.

"The challenge is to define limited war, and I would add, controlled war. Is it geographic or Is it the intensity of the operations? How big of a war is it before people start unlocking the nuclear weapons?

"Every war game I've played has seen China declare that its "no first use" policy is terminated. The China player does this to deter the United States from making precision strikes and cyber attacks on China. This seriously needs consideration before we get into a real crisis.

"Russia and China' are trying to come in with a level of intensity in escalation which is low enough so that it doesn't trigger a big Pearl Harbor response. And that could go on for a long time and is a very interesting future to explore."

Limited war requires learning about escalation control i.e. about controlled war, which when one uses that term, rather than hybrid war or gray zone conflict, connects limited war to the wider set of questions relating political objectives of the authoritarian powers.

Bracken concluded: "I believe using those terms adds to the intellectual chaos in Washington. It prevents us from having a clear policy discussion of what the alternatives for escalation control and management are in any particular crisis. This is a lot more dangerous than mishandling the Afghan exit, or the COVID pandemic."[1]

TEN

Remotely Piloted Aircraft, Autonomous Systems and How to Strengthen the ADF in the Next 3-5 Years

The Australian government is shifting resources from the Air Force and the Army and from the surface fleet to pay for a new fleet of eight SSNs with the first coming only in several years.

How then to ensure that the ADF is effective in the next five years as money and manpower is moved to what seems to be an SSN-enabled Navy with the other services adapting to this shift?

When working through the various presentations at the Williams Foundation Seminar held on April 11, 2024, there is no clear answer to this very significant challenge.

But one presentation at the seminar did raise the specter of how a pathway could be shaped to carve a way ahead, namely, the one by James Lawless entitled, "Layered Defence: The Role of Autonomy and Autonomous Systems in the Maritime Domain."

How might the new Triton Remotely Piloted Aircraft (RPA) and the various payloads which maritime and air autonomous systems deliver could accelerate change?

These systems can provide the kind of ISR and data management capabilities which Australia would need for the targeting enterprise envisaged in the "impactful projection" approach of the government which rests on effective targeting,

If one is trying to navigate the complexities of what the current Australian government is really trying to do and find a way to assess the ADF effects which result from such an effort, I would argue that one would focus on the ability to deliver strike across the areas of strategic and tactical interest to Australia and its core allies.

It is about effects and real delivery of an impact, not simply a focus on future platforms which are not going to be here any time soon.

So how to navigate through the blizzard of reports, statements and assertions by the government?

Let me start by simply citing the government's recent release indeed on their approach to strike.

According to a government press release:

Long-range strike capabilities and advanced targeting systems will receive $28 billion to $35 billion in the coming decade under the 2024 Integrated Investment Program.

The largest portion, $12 billion to $15 billion, will go to bolstering Navy's sea-based strike capability, including the acquisition of Tomahawk cruise missiles.

These will arm Hobart-class destroyers, Hunter-class frigates and, potentially, Virginia-class submarines, allowing them to hold targets at risk at longer ranges.

The funding covers Evolved Sea Sparrow Block II, SM-2 and SM-6 missiles to intercept airborne threats, along with continued integration of the Naval Strike Missile for use against heavily protected targets.

RAAF's air-launched strike capability also received investment for the F/A-18F Super Hornet, P-8A Poseidon and F-35A Lightning II to be equipped with more advanced weapons.

Funding for development of hypersonic missiles could give Super Hornets the ability to attack targets at longer ranges.

Army's acquisition of land-based long-range fires are also covered in the investment program.

This includes accelerated and expanded acquisition of 42 High Mobility Artillery Rocket Systems for Army's first long-range fires regiment.

These will fire the Precision Strike Missile that can engage potential adversaries more than 500km away.

Funding also covers Army's Guided Multiple Launch Rocket System munitions, along with new radars to extend sensor and command and control networks.[1]

But how to assess how these various programs will integrate across a kill web to deliver the kind of effects which will be credible to an adversary?

It is a question of how targeting is done, who the data for targeting can be passed to and the range of the weapon carried on a fixed or moving platform and location and with what effects when considered across the allied strike enterprise.

None of this is resolved only by funding considerations, and, for example, their needs to be a realistic public discussion of how new SSNs actually fit into the strike enterprise, for otherwise their is simply cacophony not coherence in the strike enterprise.

And any use of TLAMs by Australia in the context of a Pacific conflict where three adversarial nuclear powers are operating needs to be credibly sorted out if one is framing deterrence by denial as the core focus of Australian defence.

Malcolm Davis of ASPI raised some helpful insights in to how to interpret the government and its framing of the strike enterprise.

In his April 24, 2024 piece on "impactful projection constrained," he highlighted the following:

Strike capability featured in the 2024 update of Australia's Integrated Investment Plan (IIP), the equipment spending program that accompanied the National Defence Strategy (NDS) published on 17 April. But the strike capability acquisitions were all re-announcements—or, to take a positive view, confirmations.

They included acquisition by the navy of more than 200 Tomahawk Block IV cruise missiles, to be deployed on Hobart-class destroyers, Virginia-class submarines and maybe Hunter-class frigates. Integration of the Naval Strike Missile on surface combatants was in there, too.

The army's long-range fires mission, highlighted in the 2023 Defence Strategic Review (DSR), is centered on acquisition of 47 HIMARS launcher vehicles that can fire various long-range guided munitions, including PRsM ballistic missiles, at land and maritime targets. PRsMs have a range of 500km but could eventually reach beyond 1000km.

If forward host nation support is available in a crisis, then the littoral capability for the army will be crucial in supporting deterrence by denial with these land-based long-range fires—but we cannot assume availability of such support.

With that uncertainty in mind, establishing agreements to ensure forward host nation support for the army should be a high priority for defence diplomacy, as noted in the NDS, in coming years.

Air force capabilities include a previously announced acquisition of AGM-158C LRASM anti-ship missiles to be carried on F/A-18Fs, P-8As and eventually F-35As, as well as AGM-158B JASSM-ER air-to-ground missiles.

Another item is integration of the Kongsberg Joint Strike Missile on the F-35A. E/A-18G Growlers will get 63 AGM-88E AARGM-ER missiles for attacking radars.[2]

What is not really clear is how this fits into a strategic mosaic whereby a kill web enabled force can deliver sustained strike to provide for integrated operations in Australia's primary area of strategic interest which in my view is out to their first island chain.

This is important not just for the ADF and Australia but to credibly provide any ability to provide a sanctuary for allied forces to be able to leverage Australia's evolving support structure.

Davis went on in his article to argue for a focus on longer range strike going forward. He argued: "Impactful projection as part of deterrence by denial is the right choice—but we need to reach farther to deter more effectively. A failure to extend our reach could see deterrence by denial fall short in a real crisis."

But what remains a challenge is to build a force that would be meaningful for longer range strike which can work with allies whose interests both coincide and differ from Australia's.

What would South Korea, Japan, the United States and Australia agree on in terms of coordinated strike in a confrontation with China with North Korea and Russia almost certainly involved?

I would argue this starts by having an effective ISR integrated force which can deliver reliable data to the ADF throughout the enterprise.

Given that the government has decided to cut the fourth F-35 squadron, the RAAF is left with one significant new platform which will be crucial to shaping such an enterprise, namely the Triton.

And the introduction of Triton will be first deployed as a variety of new autonomous systems could be available to the ADF to build a layered ISR network to provide the targeting needed for both "impactful projection" and the "impactful presence".

Such capability is necessary for the direct defence of Australia and to play the role of strategic reserve for its core allies.

In fact, a layered ISR/C2 network is a key element of "impactful presence" which can be built in this interim period where the government is re-orienting the ADF in a direction towards an SSN-enabled maritime force.

Let me next turn to the presentation and discussion I had with James Lawless, the former Navy officer now with Northrop Grumman, who discussed the role which such systems can enable for Australia to have the ISR/C2 layered system which in my view is a crucial building block in the next three to five years for the ADF.

ELEVEN

Layered ISR and a Focused Force

If you want to shape an effective focused force with modest capability, for certain you need to operate as a kill web.

I focused on the concept of the kill web in a co-authored book with my colleague Ed Timperlake and then most recently in my book on the coming of maritime autonomous systems.

This is how we discussed the kill web:

In 2016, we discussed the kill web approach with Rear Adm. Manazir both when he was at N-98 and N-9 in Op Nav. With him we discussed the kill web approach as a way to shape more effective integration of forces and convergence of efforts.

The kill chain is a linear concept which is about connecting assets to deliver fire power while the kill web is about distributed operations and the ability of force packages or modular task forces to deliver force dominance in a specific area of interest.

The kill web is about building integration from the ground up so that forces can work seamlessly together through multiple networks, operating at the point of interest.

In that interview, he highlighted the key significance of evolving C2 capabilities to deliver a kill web capability.

"The hierarchical CAOC is an artifact of nearly 16 years of ground war

where we had complete air superiority; however, as we build the kill web, we need to be able to make decisions much more rapidly. As such, C2 is ubiquitous across the kill web.

"Where is information being processed? Where is knowledge being gained? Where is the human in the loop? Where can core C2 decisions best be made and what will they look like in the fluid battlespace?

"The key task is to create decision superiority. But what is the best way to achieve that in the fluid battlespace we will continue to operate in? What equipment and what systems allow me to ensure decision superiority?

"We are creating a force for distributed fleet operations. When we say distributed, we mean a fleet that is widely separated geographically, capable of extended reach.

"Importantly, if we have a network that shares vast amounts of information and creates decision superiority in various places, but then gets severed, we still need to be able to fight independently without those networks.

"This requires significant and persistent training with new technologies but also informs us about the types of technologies we need to develop and acquire in the future.

"Additionally, we need to have mission orders in place so that our fleet can operate effectively even when networks are disrupted during combat; able to operate in a modular-force approach with decisions being made at the right level of operations for combat success."

In the graphic provided by Rear Admiral (Retired) Manazir in the Williams Foundation 2018 Seminar, he took the sequence of find, fix, track, target, engage and assess and highlighted how those functions were now exercised in a distributed integrated manner by the various platforms operating within a task force or in our terms a combat cluster.

This task force, or combat cluster, can be understood either organized organically or scalable and aggregable, and operating as flexible modular task forces. With the distribution of sensors and strike throughout the battlespace, the force operates as a strike and sensing grids to gain combat dominance.

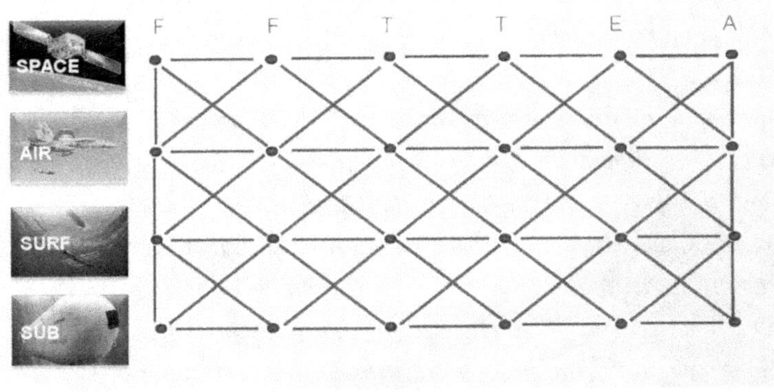

In a presentation to the Williams Foundation, Canberra, Australia on March 22, 2018, Rear Admiral Manazir then retried, provided his graphic representation of how to understand the kill web.

In some ways, the difference can be seen as a shift from a linear kill chain to a distributed kill web. The difference in focus was highlighted in a discussion in 2020 with Cmdr. Peter "Two Times Salvaggio, the head of the new Maritime Intelligence Surveillance and Reconnaissance (ISR) program at the Navy's Naval Air Warfare Development Center, at Air Station Fallon, Nevada.

He underscored: We need a paradigm shift: The Navy needs to focus on the left side of the kill chain.

The kill chain is described as find, fix, target, engage and assess. For the U.S. Navy, the weight of effort has been upon target and engage. As "Two Times" puts it: But if you cannot find, fix, or track something, you never get to target.

There is another challenge as well: in a crisis, knowing what to hit and what to avoid is crucial to crisis management. This clearly requires the kind of ISR management skills to inform the appropriate decision makers as well.

The ISR piece is particularly challenging as one operates across a multi-domain battlespace to be able to identify the best ISR information, even if it is not contained within the ISR assets within your organic task force. And the training side of this is very challenging.

That challenge might be put this way: How does one build the skills in the Navy to do what you want to do with regard to managing ISR data and deliver it in the correct but timely manner and how to get the command level to understand the absolute centrality of having such skill sets?

Here we are entering the domain of the kill web. The focus is upon how force packages are configured, and how they are empowered to leverage ISR and fire capabilities at the point of interest, and to both contribute to and leverage capabilities resident in other force packages available to deliver the desired combat or crisis management effect.

At the Williams Foundation Seminar held on April 11, 2024, James Lawless, a former Navy officer and now with Northrop Grumman, Australia, focused on how shaping layered ISR capabilities for the ADF by leveraging Triton and autonomous systems could empower the force going forward.

As Lawless underscored, that it is a daunting task to provide for direct defence of Australia given its size, location, and modest military capabilities.

He pictured the physical nature of the challenge but seen from the perspective of ISR capabilities which could be used to size the challenge and provide the manoeuvre space for the ADF to maximize their relevant impact as follows:

He argued that by building layers of ISR capability which would work seamlessly with one another, the ADF could best leverage its assets and provide decision makers with options for having the most decisive effect.

In other words, ISR is an enabling capability whereby one could winnow down the threat to determine where one needed to act and if possible, with the most decisive effect.

It was a nice to have capability: it is the indispensable capability if one is to have a focused force in reality and not just in terms of a phrase in a government document.

But then how to build this capability in the next three to five years more effectively?

The government is reducing the F-35 force by one squadron which is a major cut driven by the need to raise money for the SSN program when the government has not allocated more money to pay for it.

The major new asset coming to the RAAF is the Triton. This asset is not really understood by many defence analysts and certainly not by the public. It is a high value remotely piloted asset that can operate at high altitude and see over a wide operational area and do so why not having to operate in the weapons engagement zone.

When I visited RAAF Edinbourgh, I was impressed that the RAAF was building a common data floor for P-8s, Triton and the Peregrine. And the plan is to take integrated data and deliver it to mobile operating stations to serve the ADF.

An obvious investment which needs to be made now and capability delivered in the near to mid-term is AI enabled data management and routing to the force packages that need an integrated data stream. This is clearly a key three-to-five year capability which needs to be delivered and not just some day in the imagined world of defence procurement and (here is the killer) "planning".

But what Lawless did in his presentation was to identify various ways air and maritime autonomous systems could operate to contribute data relevant to the operations of a focused force. Autonomous systems could fill out the areas operating below Triton to move data into the weapons engagement zone as well as to inform operating forces of threats and opportunities in the battle space. They were key capabilities in terms of where on the chessboard to move your combat clusters to maximize their impact.

The Triton RPA and autonomous systems layering provide the

ADF with a significant and unique opportunity to help the government build a focused force.

But perhaps, the kill web concept might be thought of what we described as building a honeycomb deployed force when we wrote our book on rethinking military presence in the Pacific.

James Lawless presenting at the Williams Foundation Seminar April 11, 2024.

Rather than thinking of a top-down concept of managing force distribution, we focused on how you build honeycomb "cells" throughout the area of operations which could then be linked. The new technologies for ISR delivered by an RPA like Triton and the various autonomous systems which Lawless discussed can enable clusters of combat forces distributed in the area of interest or act as honeycomb cells so to speak.

In that 2013 book this is how we envisaged the C5ISR service enterprise enabling such an approach for the U.S. working with its allies in the Pacific:

"By shaping a C^{5}ISR system inextricably intertwined with platforms and assets, which can honeycomb an area of operation, an attack-and-defense enterprise can operate to deter aggressors and adversaries or to conduct successful military operations. Inherent in such an enterprise is scalability and reachback. By deploying the C^{5}ISR honeycomb, the shooters in the enterprise can reach back to each other to enable the entire grid of operation, for either defense or offense.

"In effect, what could be established from the U.S. perspective is a plug-in approach rather than a push approach to projecting power. The allies are always forward deployed; the United States does not to attempt to replicate what those allies need to do in their own defense.

"But what the United States can offer is strategic depth to those allies. At the same time if interoperability and interactive sustainability are recognized as a strategic objective of the first order, then the United States can shape a more realistic approach than one that now rests on trying to proliferate power projection platforms, when neither the money nor the numbers are there.

"In effect such an approach would be re-creating a 21st-century version of the big blue blanket. In World War II, especially in the Pacific theater, the concept of a big blue blanket evolved. It took thousands of ships and planes with appropriate logistical support to fight and win. Now with a 21st-century electronic revolution of sensors, shooters, and a honeycomb of networks a modern version of a big blue blanket can be shaped that can enable the fleet."

We then took the C^5ISR point forward into the notion of concepts of operations.

"To shape a 21st-century strategy that can encompass such challenges as dealing with the Chinese colossus, the North Korean stability and nuclear issues, the Arctic opening and the resetting of the Russian role, and providing for security for the maritime trade "highways" requires a remaking of traditional U.S. and allied capabilities and working relationships…

"The strategy is founded on having platform presence. Deploying assets such as USCG assets—for example, the National Security Cutter, USN surface platforms, Aegis, or other surface assets—and sub-surface assets, and having bases forward deployed gives the United States has core assets that if linked together into a scalable force make significant gains in capability possible.

"Such a persistent presence force must be highly interoperable with allied forces and commercial forces in order to lay down the grid that can allow for a scalable "honeycomb" of deployable capabilities.

"The honeycomb concept is central rather than simply thinking in networking terms. Various U.S. joint or allied forces can operate in an area with great autonomy, but that autonomy is not founded on significant isolation from linkage back to other forces.

"Hence the force is scalable. Scalability is the crucial glue to make such a persistent force possible. The reach from Japan to South Korea to Singapore to Australia is about how allies are reshaping their forces and working toward greater reach and capabilities.

"A scalable structure allows for an economy of force.

"Presence and engagement in various local cells of the honeycomb may well be able to deal with whatever the problem in that vector might be.

"And remembering that in the era of Black Swans, one is not certain where the next "crisis" or "engagement" might be. The author of *The Black Swan* underscored that the key impediment to learning is that we focus excessively on what we do know and that we tend to focus on the precise.

"We are not ready for the unexpected. For the author, the rare event equals uncertainty. He argued that the extreme event is the starting point in knowledge, not the reverse.

"The author in the concluding parts of his second edition advocated redundancy as a core capability necessary for the kind of agile response one needs in Black Swan or Gray Swan events. To clarify, a black swan is a large-impact and rare event beyond the realm of normal expectations. A Gray Swan is a large-impact event that is somewhat predictable but overlooked as major stakeholders in society and globally simply hope to not have to contemplate the consequences of such events.[1]

"The key conclusion here is rather simple: we need to rebuild our forces to be more agile and have more flexible expectations of what engagements we are about to engage in. And shaping plug-and-play capability with allies and partners becomes significantly more important in the period ahead."[2]

TWELVE

So What is a Multi-Domain Maritime Strategy in Shaping a Way Ahead in Australian Defence?

The seminar was concluded by comments by WGCDR Sally Knox, the moderator for the seminar, and by Air Marshal (Retired) Geoff Brown.

WGCDR Knox concluding the April 11, 2024 Williams Foundation Seminar.

WGCDR Knox provided a succinct summary of what a multi-domain approach to a maritime strategy entailed.

"An Australian Maritime strategy necessitates a highly integrated multi-agency multi-domain response enabled by, among other things, connectivity, logistics, bases, stores and decision making superiority. In short, it must be resourceful.

"Wherever given the increasingly challenging threat environment we face it must also be characterized by readiness and resilience.

"A successful maritime strategy requires seamless coordination across agencies and domains. From connectivity to logistics, every aspect must work in harmony to ensure readiness. A maritime strategy must be comprehensive.

"It isn't just about naval operations, it encompasses all aspects of national power from diplomacy to economics, safeguarding our borders and trade. Such a strategy demands a holistic approach.

"However, our demographic landscape presents a complex challenge in achieving our national objectives. Crafting a credible maritime strategy demands a multi-domain, multi-agency efforts.

"A robust maritime strategy is essential for our national security. It's not just about military capabilities, it's about leveraging all elements of power to protect our interests effectively."

Air Marshal (Retired) Brown listening to one of the presentations to Williams Foundation Seminar April 11, 2024.

Air Marshal (Retired) Brown, Chairman of the Sir Richard Williams Foundation, provided a perspective with regard to the

Defence Strategic Review and the expected Defence Investment Plan which was released shortly after the seminar.

He put his concerns this way. “Xi Jinping and Putin are 71. The strategy indicates that we are in a dangerous period something like 1936 and we have a plan for 1956. But I am not sure that these leaders are going to wait until they are 91 to do their damage.”

He then discussed the key issue which in my view really shapes the question of the credibility of the ADF going forward. He noted that the Defence Investment Plan was cutting the current force to pay for a future force, and significantly.

“Despite all the rhetoric, do we have an executable plan. How do we ensure that the ADF over the next three to five years becomes more capable?”

Or as I would put it investments for future forces to be paid for by future governments is always a tricky thing. But cuts in capability such as the fourth F-35 squadron are really and decisive reductions in the current force, the only one which adversaries see and take account of.

I also have a problem with future oriented defence planning. How well did we forecast 2020 when we were living in 2019?

THIRTEEN

April 2024 Interviews

Working the Sustainability Piece in Australian Defence: The Case of Munitions

April 10, 2024

When shaping a relevant 21st century defence approach, sustainability is a key aspect of any credible effort. Gone are the days where just in time delivery from distant global supply chains is an effective means for deployed defense assets.

Credible defense capability is built on a foundation of sustainability.

The war in Ukraine has exposed the Achilles heel of Western defense, namely the lack of magazine depth. Munitions and weapons have been in perilously short supply. Digging into one's war reserves to help the Ukrainians is short term necessity and folly.

We collectively face the challenge of building a 21st century version of the arsenal of democracy, whereby allies build munitions in common and cross support one another in a crisis. Just having a single point of failure or having to wait for delivery from a global supply chain almost certainly to be disrupted is a strategic failure of the first order.

If you are Australia, you face an especially difficult challenge as an island continent which is completely dependent in many areas on long global supply chains and a country in which manufacturing and self-processing of its rich natural resources has not been prioritized. Such a formula guarantees the absence of sustainable forces.

This situation becomes even more significant when one looks at the most plausible allied engagement strategy, namely working with all of its Pacific allies to cross-support one another, and not simply focus on the United States. By enhancing its indigenous supply capabilities, Australia can also form a strategic reserve for allies in the region or forces that might operate from Australia in the future.

But planning for such a future in the context of ongoing studies and briefing charts will not cut the cake. Briefing charts only kill the audience, not the enemy.

So what can be done in the three to five year period do achieve something real and concrete?

One answer is to build indigenous munitions capabilities, essentially a no-brainer from my point of view. If one looks at France, several years ago the government abolished the munitions facility established at the time of Louis XIV.

Just in time was enough in our peaceful world. But with Macron focusing on the need for a war economy, the French have already rebuilt their munitions production capability and are proceeding apace.

It is rather obvious that Australia needs to do the same on a priority basis.

During my April 2024 visit to Australia, I had the chance to talk with a key munitions manufacturer, Robert Nioa about the challenge.

He is head of the Nioa group which is described on their website as follows:

NIOA is a privately-owned global munitions company. Established in Queensland, Australia in 1973, today the NIOA Group has strategic locations around the world. We are dedicated to the best practice supply and manufacture of firearms, weapons and munitions to Australian and allied nation defence forces, law enforcement agencies and commercial markets.[1]

My main question to him was could they work an effective strategy of sustainable munitions supply for Australia in the timeframe which I think is critical.

According to Nioa: “Within a three-to-five-year window, we can enable Australia to provide the munitions required for an allied effort within the Indo Pacific region. We need dramatically to expand our energetics production, and we can do that within that three-to-five-year window.

“We don't have enough production capacity in Australia currently to support what we need to do for ourselves, let alone to support allies in the Indo-Pacific region.

“But we can build factories within that timeframe to provide the explosives required to produce the kinetic enablers for the ADF and as we scale up for allies in the region. We can build a factory to make solid rocket motors.

"We can build a factory to make the warheads. And then we can bring in technology for the guidance systems for long range strike or even expand conventional munitions production, everything from artillery munitions through to small arms production. It's simply an allocation of funds and priorities.

The demand signal for such expanded sustainable capability is clearly there with the shortfalls exposed in the war in Ukraine. By Australia expanding capacity they can become a strategic reserve for allies in the region as well.

And building such a sustainable infrastructure provides the material to enable lethal payloads in the future as new platforms and ways of delivering lethality evolve as well, such as I discuss in my latest book entitled *The Coming of Maritime Autonomous Systems*.

One can get caught up in imagining weapons of the future and building planning scenarios: but if you don’t have the building blocks in place for effective force sustainability, it really will not matter when you face a determined adversary that has built a sustainable force.

A 21st Century Defence Strategy for Australia: Shaping a Way Ahead

Apri 11, 2024

Many of the leading analysts with whom I have talked about the cumulative effects of the government's actions to date believe it will be to reduce the ADF combat potential for several years before there is growth in that potential. What one might call the valley of long term expectations before a climb back on the mountain of desired capability.

But what can be done to create a more effective ADF in the next 3-5 years. I really am not interested in 2030 or 2040 given our total demonstrated inability to forecast the future. I worked with Herman Kahn and Zbig Brzezinski at the beginning of my career on planning and forecasting and learned a great deal about we could not forecast. The debate about the way ahead for Australian defence as with American defence needs to shift decisively to whole of nation approaches and a much broader focus on security and defence than simply sending our troops somewhere.

It is akin to turning an aircraft carrier around at sea and heading in a different direction. And the way ahead for Australia needs to be worked in terms of compressive defence not simply the number of launch tubes at sea available someday.

But what should be the strategic re-design?

It must be built in my view on the ability to directly defend Australia out to its first island chain and to become a more credible strategic reserve for its allies in the region.

I discussed this strategic shift recently with Stephen Kuper of *Defence Connect*. The challenge of turning the aircraft carrier around and sending it in a different direction is a huge one.

And as Kuper put it: "The Australian experience is one of followership. What is needed now is a much more proactive approach to the direct defense of Australia and a 21st century one as well. Defence in the 21st century is much more than the question of lethality of your armed forces. It is about sustainability, durability, industrial capacity and the ability to mobilize the nation in defence of its interests."

We discussed the various challenges involved in a shift from followership to proactive direct defence. But throughout our discussion we focused on the seed corns for such a shift which are already in place which can be leveraged now to move further ahead.

We ended up focusing on the building of national industrial capacity to support both the ADF and allies in the region. By building out support capacities in Australia for the ADF but at the same time for allied forces in the region it was a win-win for Australia and the allies.

Kuper noted: "Too often build programs in Australia are seen primarily as jobs programs rather than national security efforts. This is both short-sided and counter-productive. The establishment of production facilities in Australia building capacity simultaneously for the ADF and allies underscores the national security dimension of what we are doing and moves us in the right direction."

His first example was of the South Korean company building a regional hub in Australia. This hub builds capability for the ADF but also serves as a strategic reserve for South Korea itself. It is a partnership not just in defence production but mutual re-enforcement of a defence partnership.

Australia needs a diversified defence partnership in the region, not simply a myopic focus on AUKUS whatever that is. AUKUS in fact is something like a Rorschach test, in that one sees one one wants to see in this vague concept.

For example, this is how the partnership has been described in an article published by the Australian government in 2022:

On 8 April 2022, Hanwha Defense Australia started construction on a new Armoured Vehicle Centre of Excellence at Avalon, Victoria. The occasion was marked by a groundbreaking ceremony attended by Hanwha Defense CEO Son Jae-il, Hanwha Defence Australia Managing Director Richard Cho, defence industry representatives, and officials from Korea and Australia.

The centre is part of a A$1 billion defence contract awarded to Hanwha Defense Australia in December 2021. It will be the first major manufacturing base for a South Korean defence company.

The centre is expected to create 300 jobs in design, engineering, manufacturing and corporate support over the life of the project.

State-of-the-facility to manufacture defence vehicles

Hanwha Defense Australia is a subsidiary of Hanwha Corporation, South Korea's largest defence company. It was formed in 2019 and is headquartered in Melbourne.

Hanwha Defense Australia will manufacture 30 self-propelled howitzers and 15 armoured ammunition resupply vehicles at the centre of excellence.

The 32,000-sqm centre will be built at Avalon Airport in Greater Geelong. The A$170 million, two-year construction will create around 100 jobs.

The facility will include:

- *multiple assembly lines*
- *a 1,500-metre test track*
- *a deep-water test facility*
- *an R&D centre*
- *an obstacle course to test capability.*

There is an opportunity for Australian defence industry partners to co-locate on site to streamline the manufacturing process.

How Austrade helped.

- *Austrade has supported Hanwha Defense Australia intensively since 2017. The agency has:*
- *shared information on Australia's defence procurement tender process, manufacturing capabilities and prospective site locations*
- *provided introductions to local Australian suppliers and R&D partners*
- *supported site visits in Australia*
- *facilitated meetings with state and territory governments.*

Austrade is seeking opportunities for Australian businesses to participate in Hanwha's global value chain. The agency will also work to facilitate exports of Australian products and services to Korea, and other international markets.[2]

The second example Kuper cited was the F-35 regional maintenance and sustainment and overhaul facilities in Australia.

He noted: "Any F 35 that's operating in the region will eventually come here to Australia for their midlife maintenance sustainment upgrades without having to go back to the U.S. So that's a big feat in and of itself. And it doesn't matter if it's an American one or Japanese one or a Singaporean one, they all can be maintained in Australia."

A January 2, 2024 *Australian Defence Magazine* discussed this development as follows:

The Government recently signed stage two of a facility services deed with BAE Systems Australia worth $110 million, which is in addition to its initial first stage commitment of $100 million announced in 2022. Minister for Defence Industry Pat Conroy said that the government's doubling of its initial investment with BAE Systems Australia will secure the Hunter Valley's future as an F-35 sustainment hub in the Indo-Pacific region.

The funding boost will enable BAE Systems Australia to build seven more maintenance bays to increase overall capacity to 13 bays to help service the growing F-35 fleet in the Indo-Pacific.

Newcastle Airport could potentially be used by other nations to sustain and service a global F-35 fleet that is expected to reach more than 3,000 aircraft.

Australian defence industry is already a vital contributor of maintenance and sustainment services for the global F-35 fleet, which is expected to reach more than 3,000 aircraft. Establishing the Hunter as an Indo-Pacific hub for F-35 repair and maintenance is a testament to the high level of skills and knowledge among our defence industry workforce.[3]

I have worked on the standup of the F-35 fleet since 2004 and have written about that enterprise in my recent book entitled: *My Fifth Generation Journey*. And I am working on a companion book which looks at the global enterprise more directly.

I suggested that such a capability would be built in Australia in an article I wrote in 2014 while in Australia attending the Williams Foundation Seminar at the time.

Kuper noted that something similar is being done with the Romeo fleet as well. As Kuper underscored: "All of Romeo maintenance and sustainment work for aircraft in the region will be done in Australia."

An Australian Department of Defence article published on June 22, 2023 highlighted this development as follows:

The first ever deep maintenance activity on a US Navy MH-60R Seahawk Romeo helicopter in Australia has been completed at Sikorsky Australia's facilities in Nowra, NSW.

The US Navy aircraft arrived in Australia in October 2022 to undergo the deep maintenance activity, known as a planned maintenance interval (PMI), which was completed this month.

Assistant Defence Minister Matt Thistlethwaite marked the milestone with Australian and US Defence and industry representatives at the Sikorsky Australia facilities on June 26.

Commodore Darren Rae, Director General Navy Aviation and Aircrew Training, said the induction of the US Navy MH-60R into Australian facilities was "a strategically significant milestone" for Australia and the United States, who share priorities to strengthen supply chain resilience in the Indo-Pacific.

"The principal aim of this activity was to demonstrate to the US the capability of Australian industry and the pathway available to perform maintenance, repair and overhaul of this helicopter in our region,□" Commodore Rae said.

"This demonstration of Australian industrial support to US Navy helicopter maintenance is a hallmark for the steady progress being made in the US-Australian alliance."

Captain William Hargreaves, US Navy H-60 Multi-Mission Helicopters Program Manager, said the activity was important in strengthening ties between the two countries and allowing Australian industry to conduct maintenance as if the aircraft was on US soil.

"Demonstrating successful PMI on a US Navy MH-60R in Australia is a testament to our two nations shared trust and commitments in our century-long partnership with the Royal Australian Navy," Captain Hargreaves said.

"Achieving this maintenance event expands our aircraft's footprint, ensuring the fleet is ready to fight tonight."

Sikorsky Australia, a Lockheed Martin company, is the Royal Australian Navy's industry partner for delivering comprehensive deep-maintenance services and intermediate-level maintenance support for the Romeo helicopter.

Commodore Rae said "Australia has a world-class industry capability in Nowra that continues to be highly successful in supporting the Royal Australian Navy's fleet of Romeo helicopters."

The Sikorsky Australia team can now add the highly successful completion of this US Navy Romeo deep maintenance activity to its list of achievements. he said.

Lockheed Martin Australia and New Zealand's Chief Executive Warren McDonald said: "Sikorsky Australia was honoured to be entrusted by the Australian-US navies as their industry partner to execute this inaugural proof-of-concept service on a US Navy MH-60R Seahawk ˜Romeo".

"Sikorsky Australia's skilled aircraft maintenance engineers have completed 46 PMIs on the Royal Australian Navy's MH-60Rs. This experience combined with the Romeo configuration alignment and interchangeable maintenance procedures between our two navies enabled seamless integration of the US Navy Romeo into Sikorsky Australia's maintenance program.

"Congratulations to the team, the Australian-US navies and Sikorsky Australia, on this exemplar PMI achievement. It underlines our shared commitment to advancing military interoperability, strengthening the MH-60R global supply chain, and showcases Australia's world-class industrial capability to sustain the most advanced maritime helicopter, the MH-60R."

The collaborative effort and dedication from teams in the Australian and US Navies have been recognised at the highest levels this year.

Australia's former MH-60R Seahawk Romeo Helicopter Assurance Program Co-Lead, Commander Andrew Newman, was presented with a Conspicuous Service Cross in January 2023 for his outstanding achievements and devotion to pursuing this activity.

Similarly, the US Navy's PMA-299 Special Project Team was awarded the NAVAIR Commander's Business Innovation award in May 2023 for their innovation, attention to detail and expertise, resulting in the first ever US Navy aviation deep maintenance activity by an Australian industrial partner.[4]

A final example we discussed was the Naval Strike Missile. This missile was designed and originally built in Norway but with its global success Konigsberg has stood up global production facilities. Australia is acquiring the NSM but by building support facilities for the ADF in Australia, they can support the region, providing strategic depth of the NSM users in the region

An August 25, 2023 article in Australian Defence Magazine highlighted Konigsberg activity in Australia as follows:

South Australian based Aerobond and Kongsberg Defence Australia have

announced a new contract for the production and provision of launcher canisters for the Kongsberg Naval Strike Missile (NSM).

The NSM was selected in 2022 to replace the Harpoon Anti-Ship Missile capability under Project Sea 1300. The contract between Aerobond and Kongsberg will result in domestically produced launcher canisters to be employed on the Royal Australian Navy's Anzac Class Frigates and Hobart Class Destroyers.

Aerobond will manufacture the NSM launcher canisters in a new 3500 square metre facility in Adelaide from January 2024, creating an additional 35 roles.

"The contract with Kongsberg is a recognition of the deep skills, experience and expertise that we have cultivated within Aerobond and across our workforce. It is an important milestone in our pathway to becoming a leading supplier in the Defence industry supply chain," said Justin Struik, Aerobond's Founder and Managing Director.

"We look forward to working with the team at Kongsberg and delivering exceptional quality to support the Royal Australian Navy and its protection of Australia."[5]

In effect, this activity can be understood in Kuper's language, "as building a mini-arsenal of democracy in the region. We have to break out of the old and antiquated view of defense whereby national security is purely things that exploit the advent of fifth generation warfare. Gray zone warfare means everything is fair game. If they're going to leverage the whole of nation against us, we must leverage the whole of nation in return."

The Future is Now for the ADF: Shaping Space for Maritime Autonomous Systems

April 15, 2024

In addition to any longer term additive capabilities, it is crucial in the evolving strategic context to find ways to enhance the ADF in the near to mid-term. This means finding ways to do so.

Clearly one way to do so is in terms of building in operational space within the operating force for autonomous systems. In my recent book on the subject, I highlighted in detail how this can be done with the extant maritime autonomous systems to provide for mission threads or specific tasks. They are not replacing crewed or manned systems but they can be delegated specific ISR and C2 tasks, and with specific ways they can be weaponized to do specific missions correlated with capital assets.

I had a chance to discuss this approach with Vice Admiral (Retired) Tim Barrett, who is not only the Williams Foundation board, but also on the Trusted Autonomous Systems Cooperative Research Centre board.[6]

As Chief of Navy, he launched the initial work on maritime autonomous systems and has seen initial systems coming to fruition. We discussed maritime autonomous systems and the way ahead with regard to the ADF.

Peter Jennings is asking a question at the Williams Foundation Seminar April 11, 2024. The author is to the left of Jennings and behind Jennings is Vice Admiral (Retried) Barrett who is about to make a comment in the seminar.

Vice Admiral (Retired) Barrett highlighted: "The surface combatant review took an eye to considering autonomous systems but considered them a generation away. But the reality is that we are already down the autonomous systems path now.

"It is wrong simply to focus on long range prospects for autonomous systems not yet here, such as platforms which could potentially carry a large number of weapons cells, rather than on the systems that are already here. The current systems can deliver significant ISR capability for example, and we need to integrate these systems into the operating force."

These systems are software and AI enabled and carry payloads. They are continuously upgraded and re-designed as they are used: they are not designed to a platform requirements standard.

As Barrett underscored: "You have to embed them into the operating force to drive the demand for further fleet innovation. They are not an add-in to some future platform.

"We need to use them actively to grow the force we need now in the threat environment we face now. We have done extensive experimentation in our Autonomous Warrior series of exercises but the future is now and we need to get on with it."

We then turned to a subject which I think highlights how you can enhance the ADF in the next three to five years with technology at hand.

I wrote a study in 2020 on the new Offshore Patrol Vessel, which is a very flexible ship being built now.[7] It is a platform designed to work with maritime autonomous systems. Given the absolute necessity to enhance maritime security in the northern waters of Australia, clearly the OPV in the hands of the Maritime Border Patrol plus autonomous systems is a way to go. And as ISR is enhanced for security purposes quite obviously that is a foundation for direct defense tasks as well.

Not only could the OPV operate as a center for managing the deployed fleet of autonomous systems but it could refuel those who needed to as well. And the crew could swap out or repair payloads on the autonomous systems. Some will be remotely piloted and those could be done from the OPV; others will be truly autonomous and directed to their tasks.

I asked Barrett about this opportunity which in my view is a low hanging fruit for ramping up ADF and Australian security and defence capabilities.

The T-38 Devil Ray is being refueled at sea by a
USCG vessel in the 5th Fleet area of operations, 2023.
Credit: MARTAC

According to Barrett: "It was the intention of the OPV to do exactly that. That is why the flight deck was retained. It was intended to compliment or supplement the hulls that are used for constabulary duty. It was to be a hull available to support the work of maritime remotes.

"But we are still experimenting. We are addressing maritime autonomous systems as if they were legacy platforms with a generational life.

"They simply are not like that. They carry payloads that are in a constant state of evolution. Their development needs to be rapid and in relation to the task at hand. They are mission thread defined: not platform defined.

"They are outside of the normal long-cycle acquisition process. In fact, the challenge is that we are NOT organized to be able to use these systems now or to engage in the transformation process driven by maritime autonomous systems.

"You cannot design a future force realistically if you are not engaged in the transformation of force through the use of maritime autonomous systems now."

The Strategic Shift in Australian Defence: Which Course to Take?

April 21, 2024

Australia as do all liberal democratic societies face significant limits on what they spend or what they can mobilize for defence.

But with the changing strategic situation in which power is diffused globally, multi-polar authoritarian movements and states are aggressively pursuing their diversified agenda, and the global "rules-based order" is not only contested but in increasing disarray, how best to shape a way ahead for Australia?

A key consideration for such a strategy is to engage the society in the defence of Australia, rather than relying on the ADF to be the sole segment of society responsible for defence.

It is also a case that government officials and strategists acting as high priests discerning what force structure one needs in the future cannot enable the society and the economy to handle the shock of accelerating global disorder.

Recently, at the Williams Foundation Seminar held on April 11, 2024, former Australian Secretary of Home Affairs, Mike Pezzullo, suggested a way ahead.

Pezzullo focused on Australian efforts to become capable of significant strides in sustainability and organizing its ready and reserve forces for a defence of Australia which would allow the island continent to become a strategic operational reserve for its Indo-Pacific allies, with the United States being central but not the only ally engaged in such an operational approach.

This is what he underscored at the seminar about considering the impact of armed conflict across the region upon Australia:

It is probable that we would face at a minimum, cyber-attacks on critical infrastructure and essential services. Cognitive warfare would be employed using technology enabled propaganda and disinformation, which will be aimed at degrading Australia's national will. We could not rule out the possibility of more direct action also being taken.

From the perspective of the United States, I would consider Australia as the vital southern bastion in any such war in relation to a number of publicly declared intelligence, surveillance, communications and space activities in support of potential U.S. operations in the South China Sea, in the South Pacific, and in the Indian Ocean, and as a haven as required for the dispersal of U.S. forces and as a logistics maintenance and sustainment support base.

If I were an American, I would also have an expectation that Australia would carry the bulk of the burden in its home theater, without unduly calling on U.S. assistance, especially where we're stretched U.S. forces to be engaged simultaneously in combat operations in Europe, the Middle East, as well as in the Indo- Pacific, if war was come to pass.

This means that defence of the Australian theatre would be a higher priority for our defence planning than sending our forces forward, except for perhaps some on a limited scale. And as our principal contribution to the war effort would, in fact, be the defence of the homeland.

As such, it would also be central to an adversary's calculations about whether, when, where to strike us and how heavily. This will have implications for how we think about force structure and how to employ it operationally.

That is we need to think of the ADF as an integrated and focused force optimized for a campaign on and around our territory, and in the broader airspace, seas and islands of the Australian theater of operations.

On April 16, 2024, I had a chance to meet with Mike Pezzullo and to discuss the way ahead. He provided an assessment of a way ahead that would meed to be embedded in Australian culture.

He started by discussing the Australian way of war. Defence has been experienced largely as an away game, and in major conflict playing a support role to a major ally.

As he explained: "We are not Israel. We are not Ukraine. We are not Finland. We have not experienced the geographical proximity of war which drives consideration for all of society mobilization considerations."

What has changed is the central significance of Australia as a location within an expanded Indo-Pacific conflict. The means of war already touch Australia daily, whether they be cyber or cognitive warfare.

Pezzullo underscored: "Not just the Americans, but our Japanese and South Korean allies look at us as the southern bastion of a comprehensive defence system to secure our interests."

The realty is simply that a 21st century defence effort rests on a significant expansion of a viable security system for the society and the economy. Much can be done with public engagement in strengthening Austrian security which also provides a more viable and sustainable defence of Australia capability.

Engagement of society in this effort in terms of steady state efforts rather than creating panic such as happened too often in the pandemic is possible. To build on public awareness of the kind of threats posed by China and other authoritarian players to Australia is an important way ahead for working ways ahead for a more resilient and sustainable Australian economy and society.

As Pezzullo emphasized: "There are several things we can do with regard to hardening telecommunications networks, actions to promoted greater energy security, shaping emergency preparations for medical care and so on that are necessary for handling emergencies and are necessary for our viability, which also prepare us for mobilization in case of war.

"We need to have ongoing discussions with sectors of the economy involved in key questions affecting continuity in a period of conflict, and this work must be realistic and precise, if not always done in public.

"How would we secure supplies in cases where shipping would be either cut off completely or significantly impaired? How would we ensure connectivity to support the war effort and to keep the central functions of society going when there's a massive attack on internet underseas cables?"

In my view, this is a shift from the high-priest vision of defence leadership to one more attune to 21st century defence needs, namely broader engagement of sectors of the economy and society in the enhanced security for the nation, which lays the foundation for more effective defence going forward.

Shaping Space for Autonomous Systems in the Operating Force: The Case of the Loyal Wingman

April 23, 2024

In the re-direction of Australian defence underway by the Labor government, a key challenge will be to make progress in the three to five years ahead for the ADF and the nation in meeting the growing global threats, while investments are being made to shape a different force down the road.

In such an effort, shaping a space for autonomous systems to assist in the process is an important aspect.

But as I have argued in my recent book on *The Coming of Maritime Autonomous Systems*, such systems complement the manned force, they don't replace.

In fact, they will be incorporated not with a replacement of platform logic but as a kill web logic: how do these platforms as payload enablers add specific capability to solve key problems facing the force?

They are not one to one platform replacements which means that the logic for their inclusion in the forces is not inherited from the past but is part of shaping an innovative way forward. What these systems provide are payloads which can assist in various tasks to augment the lethality and survivability of the manned force.

A good example of what is entailed is the Loyal Wingman program in Australia.

In a February 9, 2024 press release this is what the government said about the program:

The Albanese Government has secured hundreds of highly skilled jobs while driving innovation in Australia's local defence industry with the allocation of an additional $399 million for the ongoing development of the MQ-28A Ghost Bat.

The MQ-28A Ghost Bat, known as a Collaborative Combat Aircraft (CCA), is being developed in cooperation with Boeing Defence Australia. It is the first military combat aircraft to be designed, engineered and manufactured in Australia in more than 50 years.

An entirely new technology, it is designed to act as a loyal wingman which will be able to protect and support our military assets and pilots and undertake a wide range of activities across large distances, including performing combat roles.

The Government is now moving forward with the next stage of the program, including delivery of three Block 2 aircraft which have an enhanced design and improved capabilities. This funding boost will enable a focus on developing sensor and mission payloads, an integrated combat system and autonomous systems.

The additional funding announced today also secures over 350 jobs across Australia and will ensure ongoing work for over 200 suppliers, supporting the local defence industry and further contributing to well-paid employment opportunities for Australians.

The further development of MQ-28A Ghost Bat comes after the Government agreed with a Defence Strategic Review recommendation that options be developed for collaboration and technology sharing with the United States. In line with the Government's response, Defence signed a CCA development project arrangement with the United States on 30 March 2023.

More than 70 per cent of the MQ-28A Ghost Bat delivery program is being directed towards Australian industry content, delivering substantial benefits to local companies and their highly skilled workforces.

Quotes attributable to Minister for Defence Industry, the Hon Pat Conroy MP:

"This is the first military aircraft to be designed, engineered and manufactured in Australia in more than 50 years and underscores the depth of innovation and expertise in our defence industry.

"More than 200 Australian companies have already contributed to the MQ-28A program, including more than 50 small and medium enterprises within the supply chain. This project demonstrates that with the appropriate support from government, Australia's defence industry can continue to be a world leader and a key source of jobs.

"The prosperity and security of our nation and will always be a top priority for the Albanese Government. That's why giving our Air Force the critical capabilities it needs to protect Australians, and their interests, is paramount."[8]

Ok, but what problem is the new air system being designed to solve?

Where will it fit into the force?

And how does it add capability without burdening the sustainment enterprise within the ADF itself?

I had a chance to discuss the way ahead with Group Captain Darren Clare, Director Combat Futures. As the first squadron commander of the F-35, Clare has a good sense of the direction of air combat power, and is now working the issue of where a new autonomous air system would contribute most in the mid-term and thereby lay down capabilities for expanded use in the longer term.

A core problem facing the new generation of autonomous systems beyond the question of how to develop and build them is where to incorporate them in the force.

A loyal wingman is a key point, for a fifth-generation air force does not operate with wingmen in the legacy sense but in new ways in much wider formations. They are part of a kill web, so the place for the payloads on such an air system will be found in terms of what they can contribute that the combat force finds useful.

Where do they fit into the force?

How can they be sustained and operated by the force?

And what payloads can be credibly used by the officers charged with the authorities to use combat power for kinetic purposes?

In other words, determining where they fit into the force is crucial to ensure that we are not simply focusing on science projects, rather than on assets which are sustainable parts of an operating joint force.

For such capabilities to be significant, they must be more than aspirational: they must be in the hands of warfighters who understand what problems they are solving for the combat force.

In the discussion with GPCAPT Clare he underscored a number of key points which need to be addressed in order to bring the new systems into the operating force.

First, there is a need to demonstrate credible, achievable roles for autonomous systems.

Second, the demands placed on the operational force are significant and for the new systems to be adopted there is a need to convince commanders of new system's worth.

Third, the squadron commander needs to be able to control and provide feedback on autonomous systems in real-time.

Fourth, to progress to use, it is clear that experimentation and prototyping of autonomous systems are needed to find the best way to integrate them into the Air Force.

Above all, the importance of understanding the problem being solved with autonomous systems, and why they are better than traditional systems needs to be demonstrated to convince warfighters to use the capability and to figure how best to integrate it into the force in ways that enhance rather than degrade operational performance.

One of my favorite examples of the limits of technology not integrated is how useful the introduction of radar was to the U.S. forces in Hawaii in dealing with the Pearl Harbor attack.

An example of how to integrate new systems into an integrated force was Air Marshal Dowding and his crafting of integrated system for air defense of Britain rather than simply having a single new technology – radar – introduced to the force.[9]

It is clear at least in my view that an early credible use of such an air system would be to carry payloads which could make the new aircraft a node in a kill web, which could move data from Triton and distribute to the joint force.

Frankly, I think that weaponizing such a system is down the road but enhancing the contribution of P-8s and Tritons in terms of distributing data in the battlespace particularly with AI routing is a much nearer term capability which be valued by the operational force.

And perhaps this would happen earlier in the operational force if the new aircraft would not be managed by a squadron, but by an organization like the Surveillance and Response Group as part of their providing data to the joint force.

But getting on with use rather than conceiving of this program as if it were a traditional replacement air platform is critical if the ADF is to receive benefit from the program in the three to five year period in front of the ADF that needs to avoid a valley of reduced capability to pay for the future force.

Conceptualizing Australia's Maritime Strategy and Shaping a Government Approach

April 24, 2024

I followed up with Jennifer Parker on April 19, 2024 with regard to her presentation on how to conceptualize Australian maritime strategic interests and strategy.

We focused on how she conceptualized the strategy and the needed approach as much wider than a focus on ADF capabilities.

We discussed the need for reforming the Australian Defense Force's structure to address 21st century security challenges.

We highlighted in our discussion the importance of involving society and the economy in a broader conversation about defense and security, and the need for expedited capability acquisition to address existing gaps and emerging threats.

We finally focused on the challenge to shape a more ambitious approach to defense organization design, involving a broader societal and economic involvement to address capacity issues and maximize government capacity.

Parker started by arguing that "in Australia, we jump to the capability conversation to quickly."

She argued that any consideration of national security strategy must start with assessing Australia's critical vulnerabilities across various domains, including political warfare, cyber warfare, and space which affects its maritime interests. Maritime strategy then would be part of such an approach.

And done in this manner, Parker argues that "we need to address organizational structure and how we have organized ourselves to deal with our vulnerabilities."

Doing so will underscore the need for restructuring national agencies and departments to better address these vulnerabilities, with a focus on linking up broader considerations and authorities to do so.

This suggests or reinforces the need for a national security architecture to coordinate maritime security efforts. In such an architecture one key organizational issue to be dealt with is the lack of data sharing across fleets and departments, and the need for a central authority to address security crises.

Such a re-think then would lead to a broader engagement of the society and the economic leaders in shaping such a national strategy which would be inclusive of a maritime one.

Parker put it this way: "The third thing to do after focusing on vulnerabilities and government restructuring is to be really open with the public about what's happening, and why we need these changes."

This is a version of my own argument that simply pursuing a national security strategy in age where global security challenges are diffuse within our societies is simply continuing the role if national security decision makers as some sort of high priesthood.

The broader engagement of the society and economy is critical. Evolving defense needs rely increasingly on a security base which is not narrowly about defense. Much or perhaps most of the technologies to be mastered for defense come from the commercial sector. The flow of dual use technologies has changed from defense to the civil economy to operating the other way around.

Based on such a re-set the ADF needs to review its structure to address multiple domains of operations as it proceeds with its multi-domain integration.

We then discussed the capability issues.

How does Australia address its gaps in Navy capabilities, particularly in submarines and ships, in a more urgent manner?

Parker underscored that the Navy's capability acquisition process needs reform to be able to move faster and be more responsive to changing needs.

I mentioned my discussions with a senior U,S, Admiral who focused on their need to fill capabilities gaps with new or extant technologies, but the U.S. acquisition process simply does not allow them to do so.

The same is true of Australia, and this especially significant as contributions from autonomous systems – air and maritime – which are software driven payload carriers — become especially significant in force redesign an meeting shortfalls in the short and medium term.

Their constant redesign as use dictates a new approach whereby the users and the developers need to be working in an ongoing and continuous process of changing these systems based on real world experience.

When the Plan Jericho approach was launched by the RAAF one of the key themes identified was the need to enable software transient advantages for the force compared to an adversary. This is evident now in the coming of autonomous systems and how to include them in the force, but the challenge of how data is generated by the force and used in a whole of government maritime security and defense effort only is worsened by the coming of these systems.

There needs to be organizational change in the ADF and in whole of government in order to effectively employ these new platforms – who are not covered at all by the legacy acquisition process – to the benefit of the ADF, the Maritime Border Command and to the Australian government.

When such a re-design is pursued then the workforce problem changes as well. Parker emphasized the need to shift from a more traditional conversation of recruitment and retention to how agencies are organized for cross domain capacity.

How to enhance the efficacy and efficiency of the government cross domain to deliver the necessary decisions in the right time?

How to use the workforce more effectively and to assess the ability to deliver desired effects is even more important than managing the extant workforce to increase its numbers as the ADF seeks to expand.

How to Build a Focused Force for Australia's Current Strategic Environment?

May 2, 2024

Although the Australian Defence Strategic Review and the recent Surface Fleet Review indicated that the threat environment was deteriorating in the here and now, the major investments are being made for a force that will not arrive for a decade.

How then to enhance the ADF in the next three to five years?

Or in the words of Keirin Joyce: "The Defence Strategic Review (DSR) did not address how to give us a focused force to deal with today's threats. And the Fleet Review did not deal with how we can deal with the real threats of today with new capabilities we might affordably incorporate into today's maritime force."

Currently, WGCDR Keirin Joyce is a visiting senior fellow at ASPI. According to his bio on the Australian Strategic Policy Institute website:

WGCDR Joyce is an Australian Defence Force Academy graduate with an Honours Bachelor of Aeronautical Engineering. WGCDR Joyce has spent the last 18 years in support of the ADF Uncrewed Aerial Systems (UAS) capability including deployment to Iraq and Afghanistan.

WGCDR Joyce is a Chartered Professional Engineer, holds a Masters in Aviation Management (specialising in Human Factors), a Masters of Aerospace Engineering, a Masters in Military and Defence Studies, a Graduate Diploma in Secondary Education (Mathematics) and has researched part time as a Doctorate of Philosophy student through ADFA.

He is currently the Chief Engineer for RAAF RPAS MQ-4C Triton. Before that, WGCDR Joyce was the Australian Army UAS Sub-Program Manager responsible for all Australian Army UAS activities, including Army Drone Racing, and then the Royal Australian Air Force Remotely Piloted Aircraft Systems (RPAS) Sub-Program Manager.[10]

We started by focusing on maritime autonomous systems.

He pointed out that this technology is now a decade old, Australia has extensively experimented with this technology, and the technology can be used now.

He also cautioned that obviously not only the West is working this technology and with competitors and adversaries working with this technology, there is a clear need to know how to deal with adversary threats in this area as well, which will come only with extensive use and experience with remotely operated and autonomous system technologies.

Not only is the technology here, but much of it is also Australian and which the government can directly leverage. "With the exception of batteries and silicon, almost all of the technology necessary for an autonomous systems industry already exists in Australia."

In other words, not only does technology exist to enhance ADF capabilities in the near to midterm, but it also be part of a resilient defence structure. Joyce argued: "That gives us a clear opportunity to enhance the current force and carve out the future."

We then discussed air systems, and in particular the opportunity to leverage the coming of Triton.

He noted that "we are the only QUAD member that does not have armed persistent UAV strike. We will have the Triton which can provide the ISR and targeting data to a persistent airborne strike node, but we simply do not have armed persistent strike capability.

"This is clearly something which can be done to add to the kind of capability which the Air Force can provide to the Navy for maritime strike."

We discussed the absence of discussion as well in Australia and elsewhere for that matter of the unique quality of what NATO now has at the Sigonella base in Italy. NATO flies the Global Hawk, the U.S. now operates Triton, and the Reaper armed UAV is operating there as well. This creates the kind of combat cluster which a remotely piloted can deliver to the combat force complementary to manned air assets.

A key aspect of the impact which introducing and accelerating autonomous systems into the ADF is the workforce issue.

The ADF like virtually all Western militaries face recruitment and retention challenges. The autonomous systems are significant generators of data both ISR and C2 systems.

A civilian workforce or military cadre with enhanced flexibility on work hours and location can support such efforts. In effect, a data and an AI eco system is being generated and crafted which can be managed by innovative new ways for recruiting, developing and retaining workforce.

The final issue we discussed was acquisition.

I interviewed a U.S. Navy three-star Admiral last year, and he focused on his need to identify gaps which needed to be met and he and his staff could identify the bits and pieces which were needed which would fill the gaps.

But the organizational structure is not there for him to do that. And he emphasized that autonomous systems continuous redesign would be one way to fill gaps rather than going through the traditional lengthy and in the case of software and AI defined systems simply irrelevant acquisition process.

As Joyce underscored: "We have demonstrated we can do this. For example, in Iraq and Afghanistan we got very good at rapid acquisition to fill the gaps. Autonomous systems are almost by operational reality rapid acquisition systems, whereby the warrior works with the industrial code writer on changes he sees as possible or deems necessary.

"It is a different kind of demand driven operational development cycle: not a traditional acquisition, requirements driven approach."

My conclusion from my book on the coming of maritime autonomous systems is brutal and straightforward.

The technology has arrived, but our organizational culture and structures are not changing to be able to use it. It is a form of structural disarmament.

Australia's New Defence Strategy: The Challenge of Reshaping the ADF into a Focused Force

May 9, 2024

The Australian government has laid out its strategy of change for the ADF in last year's Defence Strategic Review (DSR) and in this year's defence investment plan and national defence strategy.

But frankly, deep cuts in current capability to play for the future force leaves many of us unsure of where the Australian government thinks it is going.

To clarify the new approach, I talked with Professor Andrew Carr to discuss the strategic shift. Carr is currently working with colleagues on going through the archives to work on how Australian governments in the recent past have crafted their defence white papers.

According to my conversation with Carr during my visit to Australia in April 2024, the key shift is from a general-purpose force to one focused on key scenarios in the Indo-Pacific region.

According to Carr: "There is a real dividing line in Australian strategy before and after the DSR. The Department of Defence has done their net assessment and have crafted several focused scenarios which guide their thinking about the way ahead. The reshaping of the force is focused on providing capabilities to deal with these select number of scenarios.

"The ADF was tasked with providing a variety of forces for a variety of missions without real focus on the region. That has now shifted to re-orienting the ADF on the region and the most likely scenarios of conflict.

"On the one hand, the ADF now is being given clear focus on what it is to prepare to do. At the same time, the government is facing the challenge of giving them the means to do so."

Of course, a major problem is that the scenarios are classified which is not very helpful in getting the kind of re-orientation needed because it is not simply an ADF tasking issue. It is about refocusing the society and economy to deal with these changes.

And other governments when making such a shift were quite able to put the scenarios driving their changes into the public domain for scrutiny.

And I think this is especially true when a government robs Peter to pay for Paul, in this case Peter being the extant ADF to pay for a future ADF.

And what is really missing is any coherent discussion of what the ADF after the acquisition of SSNs, the major driver of the shift in budget, would actually do as an integrated and coherent force in being able to deliver the deliver capability to deal with the priority scenarios.

How does the building of an SSN-enabled force improve the ADF's ability to deal with the scenarios prioritized by the government?

Not surprisingly Marcus Hellyer has provided a detailed look at the Defence Investment Plan and cut through the numbers to get to the bottom line of what the government is actually doing.

His top line analysis is as follows:

The additional Defence funding announced by the Government ($5.7 billion over the forward estimates, i.e., the next four years, and $50.3 billion over the decade) will be consumed entirely by the nuclear-powered submarine program, the general purpose frigate project and exchange rate compensation. There is no new money for anything else.

The growth in funding over the forward estimates is not unusually large by historical standards.

The new IIP repeats the failure of previous investment programs, hoping to raise acquisition spending to a wildly implausible 42% share of the total Defence budget. Defence has been trying to reach 40% since the 2016 Defence White Paper but never passed 31%.

Put another way, delivering the program requires massive, continual increases in acquisition spending. Defence is unlikely to spend this money since it has underachieved against its acquisition budget by $22.5 billion since the 2016 White Paper.

The budget contains no compensation for the loss of buying power caused by three years of extremely high inflation.

Consequently, the new program only addresses the 'exploding suitcase' of the capability program by removing large, previously planned capabilities (either completely or moving them beyond the decade). Little to no explanation is provided for these decisions.

The IIP doesn't address ADF's fundamental people program, namely that it needs to grow by 20,000 but has achieved virtually no growth over the past eight years.

Planned spending on the Maritime domain has grown from 28% to 38% of the investment budget. That's more than Land (15%), Air (14%) and Cyber (7%) combined.

No explanation is provided for why this distorted balance of investment provides better support of the new 'Strategy of Denial' than any other mix other than statements that this results in a 'focused force'.

That 38% investment in Maritime capabilities represents $114-145 billion over the decade. However, $75-95 billion of that (65%) is programmed *just for two capabilities: nuclear-powered submarines and Hunter-class frigates. Since the first of class of those fleets will only enter service at the end of the decade, two-thirds of the Maritime program's spending (and 25% of all acquisition spending) provides virtually no sovereign capability over the decade.*[11]

At the heart of this effort is a significant shift from Air Force to Navy spending, but without a coherent strategy of what exactly will the new maritime force do.

The oft used phrase of the government is "impactful projection" but as Stephen Kuper has underscored: "In order to avoid repeating history, it is clear that Australia and the ADF must begin to view expeditionary capability and the underlying doctrine, force structure, and platforms as a fundamental component of the nation's new strategic paradigm.

"Only our capacity to deploy to defend and support our regional partners and in defence of our interests through "impactful presence" will ensure that Australia's critical sea lines of communication remain unmolested in the era of great power competition."[12]

A Focused Force: Autonomous Systems and a Distributed ISR Enterprise

May 14, 2024

As the Australian government shifts the direction on building out the ADF future force, having an effective distributed ISR enterprise is a crucial element in enabling such a force.

The investments the Australian government is making in the future future force underscores the need to have accurate ISR indispensable for a distributed force. Providing coverage for the distributed operations of the ADF and in a coalition context, powerful and accurate ISR capabilities, within that distributed force, is a vital element for their survival and operations.

Fortunately, the way ahead for ISR for a distributed force is getting better due to the innovations in sensors and transmission capabilities among sensors due to progress in both commercial and defence industry.

And the coming of autonomous systems allows for the operation of ISR nests within an embedded force and the emergence of airborne AI to help shape parsimony in the distribution of relevant data to a distributed force.

I had a chance to discuss this topic with James Lawless, a former Royal Australian Officer and now with Northrop Grumman. Lawless provided a presentation at the 11 April 2024 Williams Foundation seminar focused on how autonomous systems could contribute in a major way in the near to midterm for ADF efforts to shape a distributed ISR enterprise.

It should be noted that the new capability coming to the ADF is the Triton Remotely Piloted or RPA. In my view, this platform has been viewed as simply an additive to the ADF in pursuit of advanced ASW capabilities.

But is much more than that.

It is a very high altitude aircraft with a multitude of payloads and because the Triton can operate outside of the primary weapons engagement zone, it can function as quarterback to deliver ISR throughout a very large swath of the battlespace. And, in this sense, could relay information to various types of air and maritime autonomous systems operating in support of a distributed force.

It could relay information to a loyal wingman UAV in which the wingman is supporting the attack and defense force operating in the weapons engagement zone. This RPAS could deliver and receive information to/from autonomous USVs or in certain conditions to autonomous UUVs.

These systems could, with their own AI and edge processing capabilities, mix the Triton data with their own and deliver a focused package of ISR to the combat force. The aim would be to minimize or reduce the workload of the joint force commander without any reduction in ISR data.

This is how Lawless explained the approach to me in a meeting I had with him following his presentation during my April 2024 stay in Australia.

"With the evolution of software on the Triton, there is no reason Triton can't make its own decisions about tasking other ISR assets. If we integrate autonomous AI data management capabilities on the aircraft, there is no reason Triton could not function as a quarterback for distribution of ISR data packages to autonomous platforms deployed with the distributed force."

In my work on maritime autonomous systems, I highlighted that such systems are being used to perform specific mission threads. If one focused on innovation in the ISR mission thread to build an enterprise which leveraged such systems, a key enabler for an effective distributed force is created. Because of the fifth-generation revolution built around disaggregating sensor from shooter where appropriate, this revolution continues but by using RPAs and autonomous systems in a combined arm operation.

But for this to happen, the ADF has to train differently, and procure differently. It is about culture as much it is about technology.

The example I often provide is comparing how the Americans did this with the introduction of radar in World War II with the British. The U.S. had radar at pearl harbor and even saw incoming Japanese planes. But that situation did not work out so well.

In contrast, Air Marshal Dowding put together a different type of organization into which radar was inserted and creating this air threat identification ISR and C2 system made the difference in the Battle of Britain.

We are at a similar point whereby we could create a new and effective ISR service for a distributed force. But buying bits of kit will not do it. We need full up a different organizational and training approach to grasp the future and insert it into the combat force.

A Shift in the Approach for Australian Defence Industry

May 16, 2024

To achieve the kind of resilience and sustainability which Australia requires in dealing with the threat environment, Australia needs to build more focused defence industrial capability.

What kind of capability is most needed in the evolving strategic environment?

To understand the nature of the shift required, I talked with

Professor Stephen Frühling. Professor Stephan Frühling teaches and researches at the Strategic and Defence Studies Centre of The Australian National University and has widely published on Australian defence policy, defence planning and strategy, nuclear weapons, and NATO.

He is one of the authors of recent comprehensive report precisely focusing on the nature of the shift required for Australian industry, commercial and defence, to support Australia in the new strategic environment. That report was published in December 2023 and was entitled, *Defence Industry in National Defence: Rethinking the future of Australian defence industry policy.*

The executive summary of the report is as follows:

As our geo-strategic environment deteriorates, the Australian Government has adopted the concept of National Defence – the defence against potential threats arising from major power competition – as a new approach to defence planning and strategy.

While many reforms will be required to implement the National Defence concept, building Australia's defence industry capability is one of the most important. The Defence Strategic Review has argued for the need to build enhanced sovereign defence capabilities in key areas.

However, the current paradigm of defence industry policy was established in a very different context to that of today. Risks of major power conflict were low, policy assumed a 10-year warning time, and industry capability was viewed largely in terms of supporting individual ADF programs.

This report examines the role of defence industry in the context of Australia's National Defence strategy. It argues that a change is required to recognise defence industry not as an input to capability but as national capability in its own right. The possession of a sovereign but internationally linked defence industry is itself an asset during a period where the risk of major conflict is rising.

To inform the national debate in Australia, this report examines defence industry policy in five countries: Sweden, France, the UK, Israel and Canada. These case studies offer pertinent lessons for how defence industry policy can be implemented in different strategic contexts.

The report identifies several factors that shape effective policy: fostering defence-civilian industry embeddedness; utilising a broad range of industry policy tools; ensuring formal and informal coordination between government and business; balancing competition and strategic relationships; and leveraging international markets for scale.

The report then connects these lessons to Australia, considering how our defence industry policy could be reformed to deliver on the needs of a National Defence Strategy. It offers five recommendations for the future of defence industry policy in Australia.

Policy Recommendations

- *The Australian defence industry should be considered a capability in its own right: A capability that supports the ADF force-in-being, but whose strategic value lies in those situations where that force is fully committed, needs to be rapidly reconstituted, and may need to expand. Domestic industrial capability should be developed to meet the demands of our defence planning scenarios, with foundation capabilities in place and capacity to scale with operational needs during conflict.*
- *Defence industry should be embedded within and managed as part of Australia's broader national industry structure and policy. Defence industry draws on resources such as capital, technology, infrastructure and skills from the civilian economy, and can achieve better scale and efficiencies when connected to their civilian peers. Industrial policy support for defence industry is integrated with, and not simply alongside that, support offered to its civilian counterparts.*
- *Defence industries should be strategically prioritised, then supported to achieve scale and surge capabilities. Prioritisation will be required to identify where Australia has relevant capabilities, or might be able to efficiently develop them, that can contribute to our own and allies supply chains. These capabilities should also be aligned to existing areas of strength in Australia's civilian industries and leverage new industrial policy programs. Scale in these prioritised areas should then be achieved by coordination across programs, the development of export markets, and/or the building of international technology partnerships.*

- *Government should utilise the full range of policy levers at its disposal to shape defence industry outcomes. This including both formal and informal mechanisms for coordination between government and business, to ensure greater understanding, cooperative relationships, and two-way flow of information. Given the size of Australia's defence effort, the selective use of single supplier (strategic partnering) arrangements will be crucial in some areas to achieve and sustain required industry outcomes.*
- *Government should establish a Defence Industry Capability Manager. The Capability Manager would be responsible for defining the capability and capacity that government needs to develop, as well as for development of industry to meet the level of preparedness determined by the Government. Whilst close liaison within the Department of Defence and specific Capability Managers would be required, the Industry Capability Manager would have a wider 'whole of government' role to bring Defence, wider government and industry together for the achievement of strategic industrial outcomes.*[13]

In our discussion, we focused on three main elements of fundamental change.

First, the traditional focus of Australian government's relationship with defence industry has been on platform acquisition and sustainment. The (often conflicting) aims were minimizing cost and maximizing jobs in Australia – not creating enduring industrial capacity that could scale or be leveraged to emergent needs.

And hence the government has had single platform competitions with fairly little regard for the resulting industry structure or working with a core company to craft an ongoing industrial capacity.

Frühling argued that the focus needed to shift to industrial capability and capacity, notably in what one might call the enablers of military capability, the shooters, C2 and sensors, which can be crafted, evolved and shaped by Australian industry with its own resources and in close cooperation with core partners.

Much of the technology enabling C2, sensors and even weapons comes from dynamic changes in the commercial sector and Frühling argued that there needed to be a broader focus on the industry ecosystem that could support defence in conflict, rather than on narrowly considered defence industry per se, and certainly a defence industry on a leash from government to compete platforms, choose a platform and manage sustainment of that platform.

I would add that with the shift from platforms to payloads, such a shift is absolutely crucial to shape the kind of focused but integrated force the DSR has hypothesized is necessary.

The enablers for the force are increasingly significant and the new class of systems, such as maritime autonomous systems, require a very different relationship between the users and the developers.

In security and combat operations, the use of autonomous systems drives change desired by the users who will demand code changes directly from the code writers.

As early as 2015, the RAAF articulated the need for this change and called it the capability for software transient advantage. That need is now center stage and the ADF needs a different working relationship with industry to achieve this crucial warfighting and security capability.

Second, there clearly is a need for scale in providing for enough supplies and warfighting capability in times of crisis.

Here Frühling underscored what I have called shaping an allied arsenal of democracy. Frühling underscored that Australia had limited capacity to produce platforms but by focusing on weapons, sensors and C2, it could build depth of supply in a crisis that could leverage platforms of opportunity if required – much akin to the rapid innovation we now see in Ukraine.

Hence, third, there is the importance to master systems engineering for integrating C2, ISR and weapons on a variety of platforms in times of crisis. For example, commercial vessels and novel use of autonomous platforms could be configured with C2, ISR, and weapons capabilities in times of crisis.

Frühling noted the the government strategic investment in CEA radars was an example of moving in the right direction. CEA radar modules are built into a variety of land, sea and air platforms which gives the company the scale and certainty to deliver the kind of sensing capability which the ADF needs (and they build excellent radars for export as well).

In short, the ADF needs a different kind of defence industrial ecosystem to deliver the capabilities envisaged by the DSR. It remains to be seen whether the cultural changes required to create such a defence industrial ecosystem emerge and drive such a shift.

What is the Opposite of Surge for Defence: The Australian Government Defence Strategy, 2024

May 17, 2024

When the past government and the current one underscored that the warning time for the defence of Australia has been dramatically reduced because of Chinese behavior and lack of commitment to the "rules-based order", it seemed like a critical turning point.

Only it really is not.

The current government has cut significantly current and already planned ADF capability, such as the eliminating the 4th F-35 squadron in favor of a new SSN and new surface ships in the future decade.

Not only is this not only a Paul Revere moment, it looks punting the ball moment.

And the shift being planned is dramatic.

Marcus Hellyer has carefully gone through the figures in the defence industrial plan and concluded that there is indeed a dramatic shift.

As he concluded about the plan:

It's clear we've moved on from a balanced force. But what have we moved on to? The NDS states that the ADF is now becoming a focused force. However, it's not quite clear what it is focused on doing since the NDS states (page 7) that the ADF still needs the capacity to:

- *defend Australia and our immediate region;*
- *deter through denial any potential adversary's attempt to project power against Australia through our northern approaches;*

- *protect Australia's economic connection to our region and the world;*
- *contribute with our partners to the collective security of the Indo-Pacific; and*
- *contribute with our partners to the maintenance of the global rules-based order.*

That pretty much covers every task in every part of the world which doesn't sound very focused. The NDS also says that the ADF force structure is focused on deterrence and supporting a 'Strategy of Denial'. It doesn't really say who it is trying to deny or deter from doing what. One can assume it's China, but from doing what exactly is not quite clear. Whatever it is, we seem to need significant maritime capabilities to deter China from doing it—but only at some point in the distant future.

We've noted that Maritime capabilities absorb an unprecedented 38% of the acquisition budget over the coming decade. Those are split into two main categories: Undersea warfare at $63-76 billion and Maritime capabilities for sea denial and localised sea control operations (when somebody comes up with as clunky a title as that, you know we are in the realm of deep conceptual confusion) at $51-69 billion. Those total to $114-145 billion with a mid-point at $129.5 billion.

However, only two projects dominate that spending: the SSN enterprise at $53-63 billion and the Hunter-class frigate at $22-32 billion. Again those two figures sum to $75-95 billion with a mid-point at $85 billion. Two capabilities alone consume $85 billion (remember those figures are just the spend over the decade, not the total acquisition cost). That's 65.6% of the Maritime spend. If we multiply 38% by 65.6% we can see that 25% of Defence's acquisition spend over the decade goes on just two capabilities.

But it gets worse because those projects—all going well—will only have just started to deliver actual capability by the end of the decade. The first Hunter is due to be delivered in 2032 and enter service in 2034. The first SSN is scheduled to be handed over to the RAN around 2032—all going well in Defence's most complex megaproject ever. And even then, one submarine or frigate does not a capability make.

In summary, we get virtually no in service, sovereign capability in return for 25% of Defence's acquisition spend over the coming decade—and Defence has had to give up or defer a lot of planned capability to achieve that result. What-

ever the balanced force is moving on to, it's going to take a lot of time and money to get there with little medium-term return on that investment.[14]

I would add to his argument that the actual role of the Royal Australian Navy within the Australian defence strategy is not at all clear.

When one has what is in the future the force which is programmed, just what is the con-ops of that force in the defence of Australia?

I had a chance to continue my discussion with Peter Jennings during my April visit to Australia.

And he explained why it is so difficult for an Australian government to surge rather than to plan.

As he noted: "Our professional military discuss with their counterparts, joint operations and the challenges to be met. But no government wants to discuss in public what we would actually do in a crisis. In fact, officials in government do not even wish to discuss such issues."

The government is reluctant, according to Jennings, to discuss openly realistic scenarios for joint operations. Discussing military contingencies at the political level is not an Australian political art.

I would add that failure of the political class in the West to spend time in training in crisis management is a major one. During the Cold War, I personally participated in a number of exercises with actual political leaders honing crisis management skills. This is certainly evident today.

Jennings added that for a Labour Government there is little interest in discussing military contingencies for another reason.

Australia's sovereignty will be "compromised" by working in a coalition where Australia is clearly a subordinate partner.

And with the prospect of Trump returning to the U.S. presidency and demanding "where is the beef" in current ADF capability, such a prospect is not one which the Labour Government would look forward to.

We then discussed acquisition reform difficulties.

Jennings noted that the so-called AUKUS 2 basket whereby Australia would gain access to technology which could rapidly be inserted into the ADF will not be if the traditional acquisition system holds sway. A prototype is not a capability until someone actually embeds it into the operating force.

I was quite struck by the difference between the activity of the Nordics whom I have visited over the same time as the Aussies and the activity of the current government.

This quote from the Swedish Chief of Staff says it all:

"We look at Ukraine, they are masters in using already developed civil technology to solve military problems. This is an area I look very carefully into because this is very interesting and very promising — if you also have the courage of sitting together in rooms and making sure that we understand each other, that this is what we need to solve this problem. Do you have it? Can we adjust it to something? If we start from the very beginning on a sophisticated system, it takes like 10 years. The time is not there.

"We have been working in a situation for decades, with a lot of time, no money. Now it's opposite: there is finance, but the time is limited. So for my generation of officers, it's a mental transition and change right now where we need to find a way ahead where we speed up."[15]

And that is the challenge facing Australia where the plan for the future is funded by cutting current capability.

FOURTEEN

Conclusion

The liberal democracies are at a significant cross roads in their efforts to shape defence and security efforts appropriate to the new age of multipolar authoritarianism.

Rather than a world of multi-polarity or great national power competition, a key aspect of the new historical epoch we have entered is multi-polar authoritarianism. Authoritarianism is clearly globally ascendent, but these regimes or groups do not share a common ideology or action program.

They are not in alliance, although they cooperate when convenient for their particular interests. They support splintered globalization which is when global rules exist to some extent to handle globally important exchanges but the authoritarians are not contributing political capital to maintaining the "rules-based order."

Many of these authoritarian states or groups have roots deeply inside Western democracies and through various means operate within Western societies, rather than simply being an external threat.

These means are diverse: cyber, economic investments, economic partners who advocate their economic interest, or in the case of a number of Middle Eastern states, the impact of a migration which has not been characterized by new arrivals in the West shedding their cultural or political identities from where they came.

At the same time, the liberal democracies or the West as it used to be called, is in the throes of significant self-questioning, internal debates, rejection of capitalism as practiced the last 50 years, and the emergence of disaggregated societies in each of the Western states.

These states work together on common issues, but cooperation is challenged by internal national or regional debates (in the case of Europe.)

This new era is a major challenge to the United States and its governing elites. It no longer commands a Western shaped global order. There is no great crusade as Eisenhower wrote about.

It is about national interests and winnowing commitments to the availability of resources, whether military or financial. The governing elite has not practiced or thought in terms of such discipline and the gap between the evolution of the new era and American leadership is clearly out of phase.

The war in Ukraine is not about democracy: it is about a NATO war with Russia to constrain Russian ambitions. What is the end game and how can America define and protect its interests?

We are not in with Ukraine to the end because Ukrainian and American interests are not the same. I remember the same team in Washington now promised the same to the Afghans. How did that turn out?

It is a case-by-case world in which national interests need to be carefully defined and resources calibrated to those interests.

The size and weight of the American debt and the limitations on American military power are real. We need a re-calibration of polices for the new historical era in line with our interests and a realistic understanding of our resources and capabilities, and their strengths and limits.

The key challenge facing Australia is whether or not the government and the country shape policies and a strategy to prevail in the new historical epoch.

How does Australia generate a credible deterrent strategy against a power that is their major trading partner?

How does Australia shape a national security and defence strategy which engages the nation and mobilizes its resources?

How does Australia do so while pursuing an energy strategy which simply does not tap the natural resources which Australia possesses in abundance?

How does Australia generate a credible ADF when the government is simply putting off investments in the force that would have to fight tonite?

How credible is the future force?

Do the elements of this force really integrate or are they really new platform stove pipes?

Does Australia have a credible alliance approach?

Is AUKUS really a centerpiece of a military renaissance?

How will Australia stand up to China while largely being a raw materials supplier to the country?

How realistic is the domestic understanding of what the threat from the new authoritarians is domestically? Information war is now a key domestic fact of life, and not simply an away game.

How capable is Australia of building the defence and security infrastructure it needs?

Can Australia defend itself as a sanctuary to the extent necessary to provide a strategic reserve for its Pacific allies in times of crisis?

How will Australia focus primarily on Indo-Pacific challenges and still remain engaged with at least some presence forces for other global regions?

How will Australia defend its maritime interests without shaping a significant merchant maritime capability?

In short, key questions need to be asked and answered and not only by Australia. Each of its democratic partners faces major challenges itself.

And collectively, we face a very challenging environment with a wide range of authoritarian actors with no interest on their part in providing their political and security capital to a "rules based order."

As my colleague Harald Malmgren put the challenge with regard to the United States:

"Although the U.S. has labelled this as an era of great power confrontation, the world is accustomed to such an idea throughout much of the history we have lived through.

"But it has been thought of as a binary choice, one between democracy and communism in the time of the competition with the Soviet Union and during the world war, between freedom and totalitarianism.

"We do not face such a choice currently.

"The United States has developed a powerful global military but the Russians and Chinese, to mention the two primary global competitors, are not prepared for an all out military confrontation with the U.S.

"Instead, working together, they are focused on avoiding direct confrontation while engaging in a in a multiplicity of disruptive military, political and economic activities globally which erode U.S. strength and prevent decision makers from harnessing U.S. military power in a focused confrontation with its adversaries.

"For example, Russia is using widely deployed mercenaries armed with economic tools to reduce dramatically the influence of France and the U.S. in large swathes of Africa.

"While China uses an aggressive array of bribes and threats to reshape the politics of the various sovereign nations throughout the Indo-Pacific area, the U.S. does not have in place official government institutions to counter this wide array to micro manipulations and interventions."[1]

This challenge facing the United States as described by Malmgren certainly faces all of the liberal democracies. This is not the return of the Cold War: it is a new historical age.

Although an island continent which is located at the end of the Pacific, it is now deeply enmeshed in the global change affecting the reshaping of the global system.

It has become a central player in how the Chinese way of

reshaping global power plays out, whether it wants to be or not. It is significant enough to set its own course but not powerful enough to do so unilaterally.

And it needs a defence policy that fits this period of global upheaval. In my view, it is by becoming more resilient and doing so by being interactive with and driving change amongst its allies to shape credible paths to more resilience in the face of the authoritarian powers that it provides a leadership role.

Australia's ability to enhance its ability for the direct defence of the continent while working with allies to operate as a sanctuary in times of crisis is the key way ahead. My colleague Dr. Andrew Carr has highlighted a way to think about this path to the future.

According to Carr:

In recent years much has been made of Australia's decision to acquire long-range nuclear-powered submarines, its desire for 'interchangeable' forces with America and the claim by a former Defence Minister that it would be 'inconceivable' Australia would not be involved in a conflict over Taiwan.

However, any expectation that Australia is on the verge of joining a grand military coalition to counter-balance China in Northeast Asia should be dashed after a close reading of Australia's new 2023 National Defence: Defence Strategic Review.

Sobered by the realities of China's rapid military growth and conscious that Washington is 'no longer the unipolar leader of the Indo-Pacific' as well as the limits of political and military support in the region for confrontational strategies, a new form of allied cooperation is emerging.

For Australia, this is the era of Archipelagic Deterrence. Canberra is building a secure southern bastion in order to deter and deny China's military any coercive role within the archipelagic zone from the north-eastern Indian Ocean, through maritime Southeast Asia and into the South Pacific.

At the same time, Canberra is relinquishing decades of guarded sovereignty, and allowing the United States to significantly expand its military footprint in Australia so as to better project power into Asia. Beginning with a U.S. Marine presence in 2011,

Australia is now allowing 'the rotational deployment of U.S. aircraft of all types in Australia,' establishing facilities to regularly host U.S. nuclear-powered submarines and building 'a combined logistics, sustainment and maintenance enterprise to support high-end warfighting and combined military operations in the region.'

To achieve the strategy of Archipelagic Deterrence, three big challenges Australia will have to overcome. First is the construction of deterrence systems.

Along with the political-strategic effort to establish credibility and communication, significant new military capabilities are being acquired, including an Anti-Access/Area Denial (A2/AD) shield, and the AUKUS nuclear powered submarines.

Second, Australia's strategic geography must evolve. In place of viewing Australia as a clearly delineated island nation with a defensive 'moat,' a flexible archipelagic sense of place is emerging. Australia's defence forces now seek to move seamlessly and rapidly across the littorals of Australia's northern shores and into the region, utilising a series of island-like 'nodes' and a dispersed network of bases and logistics facilities.

Third and finally, to align strategy and geography to force structure, Australian defence planning has embraced net assessment. This U.S. Cold War era strategic tool is being repurposed in an unusual way in order to break away from a problematic capability acquisition processes and create a 'focused force' for specific scenarios.

Australia's Archipelagic Deterrence strategy represents a more stable and resilient alignment of interests and capabilities between Canberra and Washington than forward-leaning alternatives. Australian leaders may talk loudly about pan-regional and global contributions; however, the enduring logic of the nation's force structure and posture has always been territorial security.

The 2023 DSR reinforces that tradition, with the Australian government insisting on the need to 'sharpen our focus, on what our interests are, and how to uphold them' in the face of the gravitational pull of attention towards the U.S.-China competition.

Recognition of this enduring thread of distinct Australian strategic interests will frustrate those hoping rich allies will help Washington to directly lift the military burden from its tired and distracted shoulders. So too for those who believe only allied boots on the ground in Taiwan can prevent its takeover.

However, Archipelagic Deterrence, with its creation of a secure southern bastion around Australia, deeply integrated with U.S. power projection into the region is likely to be better suited to the realities of distinct interests and capaci-

ties and ultimately will be more valuable to the management of major power strategic competition in the Indo-Pacific.

In short, shaping a way ahead for Australian defence in an increasingly contested Indo-Pacific region revolves around Australia's efforts in anchoring the capabilities of the liberal democracies to survive and thrive against the twenty-first-century authoritarians, with China as a key driver of change.

Additional Publications on Australian Defence

Joint by Design: The Evolution of Australian Defence Strategy

In the midst of the COVID-19 crisis, the prime minister of Australia, Scott Morrison, launched a new defense and security strategy for Australia. This strategy reset puts Australia on the path of enhanced defense capabilities.

The change represents a serious shift in its policies towards China, and in reworking alliance relationships going forward. "Joint by Design" is a book about Australia, but it is about the significant shift facing the liberal democracies in meeting the challenge of dealing with the 21st century authoritarian powers.

The strategic shift from land wars to full spectrum crisis management requires liberal democracies to have forces lethal enough, survivable enough, and agile enough to support full spectrum crisis management.

The book provides an overview of the evolution of Australian defence modernization over the past seven years, and the strategic shift underway.

Published December 21, 2020

Australia and Indo-Pacific Defence: Anchoring a Way Ahead

In Australia and Indo-Pacific Defence: Anchoring a Way Ahead, author and editor of over thirty books, Robin Laird, brings to bear his expertise on defence and security affairs to make sense of contemporary Australian international security and defence policy.

This is his third book focused on Australian defence.

It reveals the sharp mind of a person very well connected in Australian defence policy, academic and military practitioner circles. Laird has expertly sought to engage with and understand perspectives of Australian defence and security experts, many of whom are associated with the Williams Foundation, a not-for-profit Australian organisation established to advocate for the appropriate development and use of airpower, along with the other services, in defence of Australia and its interests.

This book echoes the work of the Williams Foundation which has encompassed reforms underway affecting the application not just of airpower, but also capabilities that apply to the maritime, land, space and cyber domains. It addresses the challenges of force modernisation and transformation in the context of fluctuating great power relativities (notably with the rise of an assertive and more confrontational China) in a dynamic Indo-Pacific region, at a time of significant policy initiatives affecting Australia and its place in the world.

These initiatives notably include Australia's 2023 Defence Strategic Review (DSR) and the implementation of the Australia, United Kingdom United States (AUKUS) advanced technical sharing agreement, helping Australia to acquire nuclear propulsion submarines and other advanced military capabilities.

This is an important book by a very well connected, informed and astute observer of Australia's circumstances as they pertain to defence challenges, US alliance dynamics, and technological as well as policy and political hurdles.

From the Forward by John Blaxland
Published on June 28, 2023

Australian Defence and Deterrence: A 2023 Update

This is a revised edition of the book originally published in July 2023. The original book was based on the Sir Richard Williams Foundation seminar held in March 2023 which occurred between the announcement of the new submarine program and the release of the strategic defence review at the end of April 2023.

This edition adds the work done at the second 2023 Sir Richard Williams Foundation seminar held on 27 September 2023. The first dealt with the way ahead with regard to deterrence strategy; the second dealt with the expanded role of multi-domain strike as an enabler of that strategy.

The book also includes exclusive interviews as well conducted after the first and second seminars so that original interviews with both ADF personnel and leading Australian strategists are included in the book.

The book brings together in one place a contemporary historical record of Australian thinking about the reset of their defence force and way ahead in defence.

As such, it is a unique volume.

Revised edition Published in November 2023.

All of these books can be purchased on Amazon.com in the United States and on amazon.com.au in Australia/

Notes

Introduction

1. https://operationnels.com/

1. The Impact of the Australian Political System on National Security

1. Parliament of Australia Library Briefing Book: Australia's security relationships - Australia enjoys bilateral security relationships with both New Zealand and the U.S. under ANZUS. The treaty, while not formally revoked after the U.S.- New Zealand nuclear dispute, no longer fully exists in practice. Nevertheless, a security relationship between the U.S. and New Zealand exists as members of the Five Eyes intelligence community. https://www.aph.gov.au/About_Parliament/Parliamentary_departments/Parliamentary_Library/pubs/BriefingBook47p/AustraliaSecurityRelationships#:~:text=Today%2C%20Australia%20enjoys%20bilateral%20security,longer%20fully%20exists%20in%20practice.
2. Robert Manne, The Monthly March 2006 Essays, *Little America: How John Howard changed Australia*, https://www.themonthly.com.au/monthly-essays-robert-manne-little-america-how-john-howard-has-changed-australia-184
3. Australian Labor Party, Election 2007, Policy Document, *Labor's Plan for Defence*, November 2007, https://parlinfo.aph.gov.au/parlInfo/download/library/partypol/HMWO6/upload_binary/hmwo62.pdf;fileType=application%2Fpdf#search=%22library/partypol/HMWO6%22
4. Patrick Walters, *The Making of the 2009 Defence White Paper*, Security Challenges, Vol. 5, No. 2 (Winter 2009), pp. 1-10, https://www.jstor.org/stable/26459239?seq=1
5. *Defending Australia in the Asia Pacific Century: Force 2030*, Defence White Paper 2009, Commonwealth of Australia 2009, p. 64.
6. Transcripts from the Prime Minister of Australia, Australian Government, 16 November 2011, https://pmtranscripts.pmc.gov.au/release/transcript-18272
7. Jackie Calmes, *A U.S. Marine Base for Australia Upsets China*, *The New York Times,* 16 November 2011, https://www.nytimes.com/2011/11/17/world/asia/obama-and-gillard-expand-us-australia-military-ties.html; Phillip Coorey, John Kerin, Simon Evans, "Japanese Subs on the way," *Australian Financial Review,* 9 September, 2014, https://www.afr.com/policy/foreign-affairs/japanese-subs-on-the-way-20140908-jeqdr; Kym Bergmann, "Making sense of the Japanese submarine option," *Australian Strategic Policy Institute, The Strategist,* 9 September 2014, https://www.aspistrategist.org.au/making-sense-of-the-japanese-submarine-option/
8. *Defending Australia and its National Interests*, Defence White Paper 2013, Commonwealth of Australia 2013, p. 11.

9. Ian McPhedran, "Abbott Government to spend $20B on Japanese submarines in major blow to SA's defence industry," *news.com.au*, 7 September 2014, https://www.news.com.au/national/south-australia/abbott-government-to-spend-20-billion-on-japanese-submarines-in-major-blow-to-sas-defence-industry/news-story/070a39ca785d2d8957212e91789fa4df
10. Daniel Hurst, "Tony Abbott upbeat on China trade mission despite diplomatic tensions," *The Guardian*, 3 March 2014, https://www.theguardian.com/world/2014/mar/03/tony-abbott-upbeat-on-china-trade-mission-despite-diplomatic-tensions
11. Kerry Brown, *'Fear and Greed': A closer look at Australia's China policy*, The Diplomat, 20 April 2015, https://thediplomat.com/2015/04/fear-and-greed-a-closer-look-at-australias-china-policy/
12. Nicole Brangwin, Parliamentary Library Research Paper, *Managing SEA 1000: Australia's Attack Class Submarines*, 26 February 2020, https://www.aph.gov.au/About_Parliament/Parliamentary_Departments/Parliamentary_Library/pubs/rp/rp1920/AttackClassSubmarines
13. Naomi Woodley, *Prime Minister Abbott praises Chinese president Xi Jinping's commitment to democracy, but tourism industry not convinced by FTA*, ABC News Online, 18 November 2014, https://www.abc.net.au/news/2014-11-18/praise-for-chinese-president/5898212
14. *Defence White Paper 2016*, Commonwealth of Australia 2016, pp 43-44
15. Nichole Brangwin, Parliamentary Library Research Paper, loc. cit., https://www.aph.gov.au/About_Parliament/Parliamentary_Departments/Parliamen tary_Library/pubs/rp/rp1920/AttackClassSubmarines
16. Zachary Fillingham, "Timeline: Freeze (and Thaw?) in China-Australian Relations," *Geopolitical Monitor*, 20 February 2023, https://www.geopoliticalmonitor.com/timeline-the-downward-spiral-of-china-australia-relations/
17. *Defence Strategic Update 2020*, Commonwealth of Australia, p.37
18. *Ibid.*, p. 14.
19. FRANCE24, *Were the French blindsided by the AUKUS submarine deal?*, 21 September, 2021, https://www.france24.com/en/europe/20210921-were-the-french-blind sided-by-the-aukus-submarine-deal
20. Matt Doran and Andrew Probyn, "Scott Morrison on the defensive over French submarine deal, after brief run in with Emmanuel Macron," *ABC News Online*, 31 October 2021, https://www.abc.net.au/news/2021-10-31/morrison-on-the-defensive-over-french-subs-deal-after-brief-t%C3%AAte/100583394
21. A reference to the sinking of Rainbow Warrior, codenamed Opération Satanique, a state terrorism bombing operation by the "action" branch of the French foreign intelligence agency, the Directorate-General for External Security (DGSE), carried out on 10 July 1985.
22. Quote from "Editor's Note, Dead in the Water," Chapter 1, *Australian Foreign Affairs*.
23. Matthew Knot, "Ignore the AUKUS hand-wringers, we need these subs for sea-bed battles: Navy chief," *Sydney Morning Herald*, 15 April 2023; https://www.smh.com.au/politics/federal/ignore-the-aukus-hand-wringers-we-need-these-subs-for-sea-bed-battles-navy-chief-20230414-p5d0ev.html.
24. Soli Middelby, Anna Powles, Joanne Wallis, "AUKUS an Australia's relations in

the Pacific," *East Asia Forum,* 4 November 2021, https://eastasiaforum.org/2021/11/04/aukus-and-australias-relations-in-the-pacific/

25. Soli Middelby, Anna Powles, Joanne Wallis, "AUKUS an Australia's relations in the Pacific," *East Asia Forum,* 4 November 2021, https://eastasiaforum.org/2021/11/04/aukus-and-australias-relations-in-the-pacific/
26. UK Parliament, Defence Committee Report, *UK Defence and the Indo-Pacific – Summary Report*, 24 October 2023, https://publications.parliament.uk/pa/cm5803/cmselect/cmdfence/183/summary.html
27. Margaret Reynolds, "A subservient defence policy undermines Albanese's successful first year," *Pearls and Irritations,* 22 June 2023, https://johnmenadue.com/a-subservient-defence-policy-undermines-albaneses-successful-first-year/
28. Zacchary Fillingham, *Timeline: Freeze (and Thaw?) in China-Australian Relations,* loc. cit., https://www.geopoliticalmonitor.com/timeline-the-downward-spiral-of-china-australia-relations/
29. Department of Defence Transcripts, Minister for Defence/Deputy Prime Minister Interview with Avani Dias, ABC News, 23 June 2022, https://www.minister.defence.gov.au/transcripts/2022-06-23/interview-avani-dias-abc-news
30. *Defence Strategic Review 2023*, op. cit., p. 23.
31. *Defence Strategic Review 2023*, op. cit., p. 23.
32. *Defence Strategic Review 2023* op. cit., p.18.
33. *Ibid.*, p. 45.
34. Michael Shoebridge, "New defence industry strategy: a dangerous framework that's wilfully blind," *Strategic Analysis Australia*, 2023, https://strategicanalysis.org/new-defence-industry-strategy-a-dangerous-framework-thats-wilfully-blind/
35. *Defence Strategic Review 2023*, Commonwealth of Australia, 2023, pp. 23-24
36. Institute for Integrated Economic Research-Australia, Projects, *Australia a Complacent Nation*, October 2021, https://www.jbcs.co/iieraustralia-projects
37. Department of Defence Media Release, *Release of the Defence Strategic Review*, 24 April 2023, https://www.minister.defence.gov.au/media-releases/2023-04-24/release-defence-strategic-review
38. Andrew Green, "Peter Dutton warns against UK submarines for AUKUS, drawing fire from the government," ABC News Online, 1 March 2023, https://www.abc.net.au/news/2023-03-01/peter-dutton-aukus-submarines-government-labels-irresponsible/102040234
39. Australian Foreign Affairs, "Dead in the Water: The AUKUS Delusion," AFA20, February 2024, https://www.australianforeignaffairs.com/
40. Hugh White, "Dead in the Water," *Australian Foreign Affairs*, February 2024, https://www.australianforeignaffairs.com/essay/2024/02/dead-in-the-water
41. Matthew Knott, "Defence Force under 'stress' as chief reveals true extent of staff crisis," *The Sydney Morning Herald,* 14 February 2024, https://www.smh.com.au/politics/federal/defence-force-under-stress-as-chief-reveals-true-extent-of-staff-crisis-20240214-p5f4xp.html.
42. John Hewson, "The path of political delusion," *The Saturday Paper,* 30 March 2024, https://www.thesaturdaypaper.com.au/comment/topic/2024/03/30/the-path-political-delusion
43. National Museum of Australia, Defining Moments: ANZUS Treaty, https://www.nma.gov.au/defining-moments/resources/anzus-treaty,

44. Ingeborg van Teeseling, *History Vietnam War, Australia Explained*, https://australia-explained.com.au/history/vietnam-war
45. "Australian Involvement in the Iraq War," *Wikipedia*, https://en.wikipedia.org/wiki/Australian_involvement_in_the_Iraq_War
46. Australian War Memorial, "Deaths as a result of service with Australian units," https://www.awm.gov.au/articles/encyclopedia/war_casualties.
47. Department of Health, Australian Government, "Life in Mind, Veterans and Australian Defence Force (ADF) personnel," https://lifeinmind.org.au/suicide-prevention/priority-populations/veterans-and-australian-defence-force-personnel
48. For example, "Department of Defence Transcripts, Minister for Defence/Deputy Prime Minister Television Interview," *Today Show*, 21 February 2024, https://www.minister.defence.gov.au/transcripts/2024-02-21/television-interview-today-show
49. Institute for Integrated Economic Research-Australia, Projects, *Australia a Complacent Nation*, October 2021, https://www.jbcs.co/iieraustralia-projects
50. Lili Bayer, "Europe must get ready for looming war, Donald Tusk warns," *The Guardian*, 30 March 2024, https://www.theguardian.com/world/2024/mar/29/europe-must-get-ready-for-looming-war-donald-tusk-warns

2. Shaping a Way Ahead for Australian Defence and Deterrence

1. *National Defence Strategic Review* (Australian Government), 2023, 30.
2. Robbin F. Laird and Edward Timperlake, *A Maritime Kill Web Force in the Making: Deterrence and Warfighting in the 21st Century* (2021), pages 41-43, Kindle Edition.
3. https://thehill.com/opinion/national-security/543023-the-risk-of-new-military-technologies-must-be-properly-assessed/. Also, see https://defense.info/re-think ing-strategy/2021/03/conventional-forces-and-technology-what-is-their-impact-on-escalation-management-with-nuclear-powers/
4. https://defense.info/re-thinking-strategy/2020/09/defending-south-of-australias-first-island-chain/
5. https://sldinfo.com/2019/05/australian-sovereignty-and-maritime-security-radm-goddard-discusses-the-role-of-the-maritime-border-command/
6. https://sldinfo.com/2023/05/maritime-autonomous-systems-providing-mission-threads-for-australian-defence-and-security-the-case-of-the-bluebottle-usv/
7. https://www.aspistrategist.org.au/turning-the-defence-review-into-action-will-require-a-major-mobilisation/
8. https://theforge.defence.gov.au/publications/can-australians-fight
9. Stephan Frühling & Andrew O'Neil, "Alliances and Nuclear Risk: Strengthening US Extended Deterrence" *Survival* (4 February 2022, p. 92).

3. The Case of Australian Maritime Strategy

1. https://defense.info/re-thinking-strategy/2021/12/gray-zones-or-limited-war/

4. General Considerations for an Effective Strategy

1. https://www.theaustralian.com.au/subscribe/news/1/?sourceCode=TAWEB_WRE170_a_GGL&dest=https%3A%2F%2Fwww.theaustralian.com.au%2Finquirer%2Fplayers-scramble-to-regain-upper-hand-on-deterrence%2Fnews-story%2Fdc4e9e5d44dcd732c9b9978f12f0844f&memtype=anonymous&mode=premium&v21=GROUPA-Segment-2-NOSCORE&V21spcbehaviour=append

5. Air Power in Australia's Maritime Strategy

1. https://defense.info/partners-corner/2024/03/raaf-p-8a-poseidon-works-with-the-french-on-reunion/
2. https://sldinfo.com/2023/04/agile-basing-and-endurability-as-a-key-deterrent-capability-a-conversation-with-the-air-commander-australia/

6. Multi-Domain Operations in the Maritime Domain: The Significance of Digital Interoperability

1. https://www.usni.org/magazines/proceedings/2012/january/long-reach-aegis
2. https://sldinfo.com/2016/09/the-network-as-a-weapon-system-the-perspective-of-rear-admiral-mayer-commander-australian-fleet/
3. https://sldinfo.com/2018/08/visiting-hmas-hobart-a-key-building-block-in-the-remaking-of-the-royal-australian-navy/

7. Multi-Domain Operations in Australia's Maritime Strategy: The Army, Navy, and Air Force Orient Their Efforts

1. https://news.usni.org/2023/10/02/australian-army-shifting-priorities-to-amphibious-littoral-operations
2. https://www.minister.defence.gov.au/media-releases/2017-10-03/new-approach-naval-combat-systems

8. Key Dimensions Required for a Successful Multi-Domain Approach

1. https://www.aspistrategist.org.au/national-defence-strategy-tackling-problems-not-just-declaring-principles/

9. Cognitive and Information War and the "Gray Zone"

1. https://defense.info/re-thinking-strategy/2021/12/gray-zones-or-limited-war/

10. Remotely Piloted Aircraft, Autonomous Systems and How to Strengthen the ADF in the Next 3-5 Years

1. https://www.defence.gov.au/news-events/news/2024-04-23/boost-ability-strike-afar

2. https://www.aspistrategist.org.au/defence-strategic-review-impactful-projection-constrained/

11. Layered ISR and a Focused Force

1. Nassim Nicholas Taleb, *The Black Swan: The Impact of the Highly Improbable* (Penguin), 2008.
2. Robbin Laird, Edward Timperlake, and Richard Weitz. *Rebuilding American Military Power in the Pacific: A 21st-Century Strategy* (Praeger Security International). ABC-CLIO. Kindle Edition, 2013.

13. April 2024 Interviews

1. https://www.nioa.com.au/
2. https://www.globalaustralia.gov.au/news-and-resources/news-items/hanwha-defense-starts-construction-new-armoured-vehicle-centre-excellence-victoria
3. https://www.australiandefence.com.au/defence/air/williamtown-f-35-facility-to-expand#:~:text=All%2072%20of%20Australia's%20F,upgraded%20at%20the%20Hunter%20facility.
4. https://www.defence.gov.au/news-events/news/2023-06-27/work-us-navy-heli copter-breaks-ground#:~:text=The%20first%20ever%20deep%20maintenance,Australia's%20facilities%20in%20Nowra%2C%20NSW.
5. https://www.australiandefence.com.au/defence/sea/aerobond-secures-manu facturing-contract-for-naval-strike-missile-program
6. https://tasdcrc.com.au/
7. https://sldinfo.com/2020/04/a-small-ship-but-a-significant-shift/
8. https://www.minister.defence.gov.au/media-releases/2024-02-09/albanese-government-invests-further-400-million-next-generation-loyal-wingman-drone
9. https://sldinfo.com/2015/10/the-importance-of-innovation-by-design-for-combat-success-building-on-the-legacy-of-air-marshal-dowding/
10. https://www.aspi.org.au/bio/keirin-joyce
11. https://strategicanalysis.org/unpacking-the-numbers-in-defences-new-inte grated-investment-plan/
12. https://www.defenceconnect.com.au/geopolitics-and-policy/13992-with-nds-and-iip-out-of-the-way-were-still-no-closer-to-understanding-what-is-meant-by-impactful-projection
13. https://www.aigroup.com.au/globalassets/news/reports/2023/ai-group-sdsc-dind-report.pdf

14. https://strategicanalysis.org/unpacking-the-numbers-in-defences-new-integrated-investment-plan/
15. https://breakingdefense.com/2024/05/swedens-top-officer-on-the-mental-transition-of-joining-nato-and-russian-concerns/

14. Conclusion

1. https://defense.info/featured-story/2024/05/america-in-2024-challenges-in-reshaping-its-global-strategy/

About the Author

Dr. Robbin F. Laird is a long-time analyst of global defense issues. He has worked in the U.S. government and several think tanks, including the Center for Naval Analyses and the Institute for Defense Analyses.

He is a frequent op-ed contributor to the defense press, and he has written several books on international security issues.

He is the editor of two websites, *Second Line of Defense* and *Defense.info*.

He is a member of the Board of Contributors of *Breaking Defense* and publishes there on a regular basis.

He is a research fellow with The Sir Richard Williams Foundation.

He is also based in Paris, France, and he regularly travels throughout Europe and conducts interviews with leading policymakers in the region.

About the Contributors

Air Vice-Marshal John Blackburn AO (Retd)

He is a former fighter pilot, test pilot, capability and strategy planner, and finally the Deputy Chief of the Royal Australian Air Force in which he served a total of 43 years.

In 2007 he was appointed an Officer in the Military Division of the Order of Australia.

He holds a Master of Arts and a Master of Defence Studies and is a life member of the Society of Experimental Test Pilots. He is the co-founder and Chair of the Institute for Integrated Economic Research Australia, a co-founder and Executive Committee member of the Australian Security Leaders Climate Group, and an editorial advisor to the French Operational SLDS (Logistics Defense Security Support) magazine.

He was formerly a Council Member of the Australian Strategic Policy Institute, Chairman of the Kokoda Foundation, Deputy Chairman of the Sir Richard Williams Foundation, and an Adjunct Professor at Edith Cowan University in Western Australia.

He also has had over 15 years experience as a consultant in the fields of Defence and National Security.

Group Captain Anne Borzycki (Retd)

She has served a total of 35 years in the Royal Australian Air Force. During her RAAF career, Anne was seconded to Parliament House to work as the defence adviser to the Joint Standing Committee on Foreign Affairs, Defence and Trade, was a liaison officer to Capability Development Group, lead a remuneration case to change the system of the payment of flying allowance and undertook public affairs and media management roles.

She is also a co-founder and Director of the Institute for Integrated Economic Research Australia, a Fellow of the Australian Security Leaders Climate Group, and a former Board Member of the Sir Richard Williams Foundation.

www.ingramcontent.com/pod-product-compliance
Lightning Source LLC
LaVergne TN
LVHW020712110826
845149LV00012B/2227

* 9 7 9 8 9 9 0 3 8 4 4 2 2 *